WATER POLLUTION BIOLOGY

**UNIVERSITY OF GLAMORGAN
LEARNING RESOURCES CENTRE**

Pontypridd, Mid Glamorgan, CF37 1DL
Telephone: Pontypridd (01443) 482626

Books are to be returned on or before the last date below

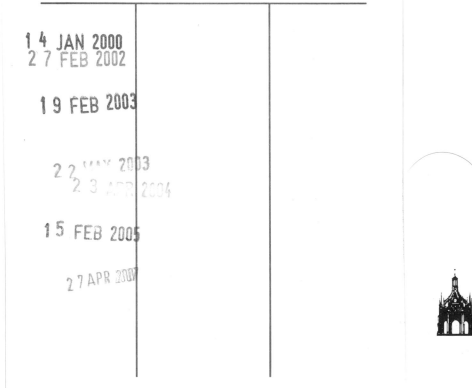

WATER POLLUTION BIOLOGY

P. D. ABEL, B.Sc.(Hons.), Ph.D.
Department of Biology
Sunderland Polytechnic

ELLIS HORWOOD LIMITED
Publishers · Chichester

Halsted Press: a division of
JOHN WILEY & SONS
New York · Chichester · Brisbane · Toronto

First published in 1989 by
ELLIS HORWOOD LIMITED
Market Cross House, Cooper Street,
Chichester, West Sussex, PO19 1EB, England
The publisher's colophon is reproduced from James Gillison's drawing of the ancient Market Cross, Chichester.

Distributors:

Australia and New Zealand:
JACARANDA WILEY LIMITED
GPO Box 859, Brisbane, Queensland 4001, Australia

Canada:
JOHN WILEY & SONS CANADA LIMITED
22 Worcester Road, Rexdale, Ontario, Canada

Europe and Africa:
JOHN WILEY & SONS LIMITED
Baffins Lane, Chichester, West Sussex, England

North and South America and the rest of the world:
Halsted Press: a division of
JOHN WILEY & SONS
605 Third Avenue, New York, NY 10158, USA

South-East Asia
JOHN WILEY & SONS (SEA) PTE LIMITED
37 Jalan Pemimpin # 05–04
Block B, Union Industrial Building, Singapore 2057

Indian Subcontinent
WILEY EASTERN LIMITED
4835/24 Ansari Road
Daryaganj, New Delhi 110002, India

© 1989 P. D. Abel/Ellis Horwood Limited

British Library Cataloguing in Publication Data
Abel, P. D. (Philip David), *1951–*
Water pollution biology.
1. Water. Pollution. Biological aspects
I. Title
628.1′68

Library of Congress Card No. 88–38863

ISBN 0–7458–0512–4 (Ellis Horwood Limited)
ISBN 0–470–21394–9 (Halsted Press)

Typeset in Times by Ellis Horwood Limited
Printed in Great Britain by Hartnolls, Bodmin

8.6.92

Table of contents

Preface

Twenty years ago, most ecologists — at least those working in academic positions — were little interested in the study of polluted environments or in pollution as a phenomenon. With some notable exceptions, they took the view that pollution was an extraneous factor which interfered with their efforts to study and understand the processes which occur in the natural environment. This attitude was strange, because in other areas of biology, such as biochemistry or physiology, the insights which could be gained into the function of normal systems by studying experimentally disturbed systems had long been recognised. Within a few years, this situation changed. It became fashionable for scientists to show that their work was 'applied' rather than 'pure', as if the distinction had any meaning. Increasing public concern over environmental issues led to the increased availability of research funding; courses at universities, polytechnics and colleges proliferated; new journals appeared to accommodate the increasing volume of published work.

The result of this sudden influx of interest and resources into the field of water pollution research was in some ways comparable to the accelerated eutrophication which occurs when nutrients are added to a lake: the increased productivity can have consequences which are not unequivocally beneficial. The academic community remained largely ignorant of the work which had already been done, over many decades, by scientists outside their community who are responsible for the actual management of the environment. They worked in Government laboratories, or in the various regulatory agencies which most people have never heard of, and published their work in specialised reports or journals which were not widely read by academics. Therefore, while many real advances have been made in the last two decades, it is at least arguable that the increase in knowledge has not been altogether in proportion to the burgeoning of the literature.

I began to write this book, after several years teaching water pollution at undergraduate and postgraduate levels, because students clearly needed help in assessing the quality of the material they were studying, and in distinguishing the sound from the methodologically-flawed, the genuine advances in understanding from the merely repetitive or derivative accumulation of data, the critical application of sound principles from speculation based on inadequate information. These distinctions are important in all branches of science, but pollution suffers more than most topics because it is, more than most, in the public domain and more subject to comment and speculation from the ill-informed. It is also, by its very nature, a topic which requires an interdisciplinary approach. There is therefore a need for specialists from different disciplines to understand something of one another's viewpoint, and to be able to explain their subjects to colleagues who have different technical backgrounds or professional experience.

In this book, therefore, I have attempted not only to impart factual information, but also some idea of the areas of uncertainty which must remain; to impart to the reader some sense of how he may evaluate for himself the quality and reliability of the vast amount of material he may encounter; and to concentrate on the elucidation of principles rather than the promulgation of data. Where appropriate, however, I have referred to the most up-to-date summaries I have been able to locate. I have assumed some previous knowledge of general ecological principles and of the ecology of aquatic habitats, but in certain places have felt it advisable to offer some summative explanations. Some readers may find these unnecessary or over-simplified, others will perhaps find them inadequate. I can only hope that for most readers the balance is about right. I have treated some aspects of the subject in more detail than others. I have, for example, covered the toxicology of water pollutants at length, because it is a subject which is widely misunderstood, even abused, and it has not been treated adequately in a text at this level before. Most of the book is about the effects of pollution on the fauna and flora of fresh waters, but I have addressed, for example in Chapters 5 and 6, some of the wider aspects of water pollution with which biologists could and should be concerned. Chapter 7 is no more than a selective introduction to a subject which requires a book to itself.

I have to thank several of my colleagues at Sunderland for their informative discussions, for their forbearance when my preoccupation with writing imposed extra burdens upon them, and for the loan of reference materials. Dr B. Mitchell and Dr P. Wyn-Jones commented upon a draft of Chapter 5; I have not always chosen to follow their advice, and accept responsibility for any errors which remain. I thank several of my students and former students, especially Dr David Green. I have made extensive use of Dr Green's study on the Rivers East and West Allen in Chapters 1 and 3. Owing to pressure of other work, he has been unable to prepare a complete account of this study for publication and it is presented here, in summary, for the first time in a widely accessible form. Mr Gilbert Wright drew most of the diagrams, but Mr E. Gilhooley, Dr R. Morrison and Dr J. Terry were responsible for some. Mrs S. Cottam and Mrs S. Tunney typed the manuscript. Dr A. Velegraki-Abel's contribution to the work at every stage was as comprehensive as it was characteristically unorthodox.

I gratefully acknowledge permission of the following to reproduce copyright material: Dr D. W. J. Green (Figs 1.1, 1.2, 1.3, 1.4, 3.15, 3.16); The Zoological Society of London (Fig. 1.6); *Nature* (Macmillan Magazines Ltd) (Fig. 2.9); Dr J. M. Hellawell and the Natural Environment Research Council (Figs 3.1, 3.3, 3.5, 3.8, 3.10); Freshwater Biology (Figs 4.5, 4.11, 4.12, 4.13); The American Fisheries Society (Fig. 4.14); the *Journal of the Fisheries Research Board of Canada* (Fig.

4.15); Pergamon Press (Figs 4.8, 4.16, 4.17, 4.18); Mr R. Lloyd and the Royal Society of London (Fig. 4.19); Mr T. E. Tooby (Fig. 6.5).

Sunderland, 1988   P. D. Abel

1

Some studies of polluted rivers

Water is one of our most important natural resources, and there are many conflicting demands upon it. Skilful management of our water bodies is required if they are to be used for such diverse purposes as domestic and industrial supply, crop irrigation, transport, recreation, sport and commercial fisheries, power generation, land drainage and flood protection, and waste disposal. An important objective of most water management programmes is the preservation of aquatic life, partly as an end in itself and partly because water which sustains a rich and diverse fauna and flora is more likely to be useful to us, and less likely to be a hazard to our health, than one which is not so endowed. To meet this objective, it is necessary to maintain within certain limits factors such as water depth and flow regimes, temperature, turbidity and substratum characteristics, and the many parameters which contribute to the chemical quality of the water. In waters which receive waste discharges, whether by design or by accident, one or more of the these variables may come to lie outside the limits which can be tolerated by one or more of the species which live there, and consequently the biological characteristics of the water are altered.

The biologist's role in the monitoring and control of water pollution is to detect and accurately to describe these alterations, to elucidate as precisely as possible the mechanisms by which they are brought about, and to seek to understand the qualitative and quantitative relationships between pollution and its biological consequences. He may also need to be aware of the application of biological processes in the control or amelioration of pollution, and of the serious consequences for public health which water pollution threatens. Finally, he must be able to offer constructive advice to other specialists — chemists, engineers, administrators and legislators — who share the responsibility for managing our water resources. All of these topics will be discussed in the following chapters, but perhaps the best introduction to the scope of the problems encountered by the biologist interested in the effects of water pollution is to look at some polluted rivers, at some methods of studying them, and at some applications of those methods in the management of polluted waters.

1.1 METALLIFEROUS RIVERS

In many parts of the world, rivers have become contaminated with heavy metals such as zinc, lead and copper as a result of mining and associated activities. In Britain, three areas of the country are particularly well known because they contain a number of metalliferous rivers: one is in the south-west of England, one is in West Wales and one is in the North Pennine Orefield in northern England. Although these areas were once major centres for the extraction and processing of metal and mineral ores, within the last hundred years or so the mining activities have declined considerably, or ceased altogether. These areas never became major population centres, and were never subject to the intensive pollution pressures of modern urban and industrial society. The effects of the mining activity are still detectable, but these effects have not been compounded and complicated by those of other forms of pollution associated with modern development. For this reason, they offer the opportunity to investigate the impact of an important group of industrial pollutants — heavy metals — on waters which are relatively unaffected by the pressures which have more recently been unleashed on water bodies in more densely-populated areas. They have therefore been extensively studied, and provide a good illustration of the biological effects of pollution, of the methods which may be used in studying these effects, and of the questions which biologists seek to investigate and answer about polluted waters. The following account is largely based upon the review of Green (1984). Mance (1987) also discusses some studies of metalliferous rivers.

Ecological studies on metalliferous rivers in West Wales began about seventy years ago, and have continued, intermittently, ever since, Carpenter (1922, 1924) observed that fewer invertebrate species occurred in these rivers at stations close to the lead mines, and that the differences in abundance of invertebrates appeared to be due to lead in the mine effluents rather than to physical differences in the river bed. Certain invertebrate groups (Platyhelminthes, Mollusca, Crustacea, Oligochaeta and many Insecta) were always absent from the most heavily polluted sites, but some insect species such as *Cloeon simile* (Ephemeroptera), *Simulium latipes* (Diptera) and *Velia currens* (Hemiptera) appeared to be tolerant of elevated lead concentrations. Following the closure of a mine and the cessation of pollution, Carpenter (1926) described a process of recovery. The first stage was the establishment of a restricted fauna consisting almost entirely of insect larvae on a substratum bearing a limited covering of algae and bryophytes. Oligochaetes, molluscs, platyhelminthes, crustacea and many insect species remained absent. Subsequently, the encroachment of chlorophyceous algae was accompanied by an increase in invertebrate species diversity, with oligochaetes, turbellarians, caddis larvae and other insects becoming established. In the final stage of recovery, macrophytes were established in physically-suitable areas and there was a large increase in the numbers of invertebrate taxa present, including molluscs, and the development of fish populations. This process of recovery which occurred following the closure of a mine could also be observed in reverse (Carpenter, 1926) when a mine which had been closed for some years was subsequently re-opened. The same stages of succession could also be observed contemporaneously in successive reaches of a polluted river with increasing distance from the source of pollution (Carpenter, 1924). The toxic agent responsible for these changes was at first assumed to be lead, particularly dissolved lead rather than the particulate component of the mine effluents. Carpenter (1925) investigated the toxicity of dissolved lead salts to fish, but at the concentrations of lead found in the river water, lead salts alone were less toxic than the river water. This indicated that the mine effluents contained some additional toxic agent, and zinc was soon identified as being an important factor.

During the decades following Carpenter's early investigations, advances in techniques of taxonomy, ecological sampling and chemical analysis provided more detailed information on the effects of metal pollution in upland rivers. Jones (1940a, 1940b) studied the Rivers Ystwyth and Rheidol, also in Wales and confirmed the general pattern described by Carpenter. In the Ystwyth it became clear that zinc, rather than lead, was the most abundant toxic agent; by 1958 lead levels had declined markedly but zinc levels remained high (Jones, 1958). This probably occurred because zinc is more soluble than lead, and because the pattern of mining activity in the area tended to result in the fairly efficient removal of lead and silver from the ores, whereas waste material dumped in and around the mines remained rich in zinc. Between 1940 and 1958, there was no substantial increase in the number of species found in the Ystwyth. Molluscs, crustaceans, oligochaetes and leeches were still absent. Among the insects, the fauna remained restricted but *Rhithrogena semicolorata* (Ephemeroptera), *Simulium* spp. (Diptera), and the stoneflies *Leuctra* spp. and *Nemoura* spp. were fairly numerous. Caseless caddis larvae were more numerous than cased caddis larvae. On the basis of some toxicity tests, the absence of fish, molluscs and crustaceans was ascribed to the presence of toxic metals in the water, but the scarcity of flatworms, and the absence of leeches and oligochaetes was thought to be due to the unfavourable characteristics of the substratum. By 1980, little further improvement had occurred. Brooker & Morris (1980) recorded more species than were found in 1958, but much of the increase is apparently due to the fact that some groups of animals earlier identified to genus, particularly chironomids and simulids, were identified to the species. Paradoxically, the association of particular metal levels with the biological characteristics of the water is becoming more difficult. For example, it is now known that high levels of calcium to some extent protect animals from the toxic effects of heavy metals, whereas low levels can, quite apart from any influence of heavy metals, themselves act as limiting factors. A great deal more is now known also about the influence of purely physical factors on the patterns of distribution of invertebrate species. Therefore it is less easy than it might earlier have appeared to interpret the available biological data solely in terms of the measured heavy metal levels. Nevertheless in the nearby river Rheidol, where metal levels are lower, the invertebrate community is distinctly richer. Whereas Carpenter (1924) recorded 29 species, Laurie & Jones (1938) recorded 103 following a distinct reduction in the metal levels some years earlier; Jones (1949) recorded 130 species, though molluscs and crustaceans were still absent. Brooker & Morris (1980) recorded 134 species.

Elsewhere in Britain and in other parts of the world, metalliferous rivers similar to those in Wales have been widely studied, though rarely over such a long period of time. In south-west England, Brown (1977) studied the River Hayle which is contaminated with zinc, copper and iron. Metalliferous rivers in the North Pennine Orefield have received some attention (e.g. Armitage, 1980). Examples from North America include the studies of Sprague *et al.* (1965) on rivers polluted with copper and zinc; and of Gale *et al.* (1973) on rivers polluted with zinc and lead. Zinc-polluted rivers in Australia have been investigated by Weatherley *et al.* (1967). The river South Esk in Tasmania, which contains cadmium, zinc, cooper, iron and manganese has been investigated by Thorp & Lake (1973), Tyler & Buckney (1973) and by Norris *et al.* (1982). There are, of course, many other rivers polluted with heavy metals, but in most of these the effects of heavy metal pollution are compounded or modified by substantial inputs of organic matter and/or other industrial pollutants. Based on his review of investigations relating to rivers polluted solely or predominantly by heavy metals without these other complicating factors, Green (1984) drew attention to the following general points.

In rivers polluted by heavy metals, the invertebrate fauna is affected by the elimination, or numerical reduction, of certain species. If the input of pollution ceases, the invertebrate fauna gradually recovers with the passage of time. A similar process of recovery can be observed with increasing distance from a source of pollution. The nature of the recovery is a gradual increase in the numbers of species and in the numbers of individuals found in the water. Taxa which are universally affected by metal mining and associated activities are Mollusca, Crustacea, Platyhelminthes and Oligochaetes. Some groups, however, appear to behave inconsistently in response to metal pollution. One such group is the larvae of caddis flies (Trichoptera). In the Rheidol and Ystwyth, for example, Jones (1940b) noted the absence of cased caddis larvae from polluted stations, whereas carnivorous, caseless species such as *Rhyacophila dorsalis* and *Polycentropus flavomaculatus* were present. In laboratory experiments, no evidence was found that the concentrations of dissolved metal found in the rivers was lethal to cased caddis species, and in fact some species were able to construct their cases, in the laboratory, from particles of solid mine waste. This suggested that metals may exert indirect effects on some species in the field, possibly related to the effects of the metal on the food source of the caddis larvae. An alternative explanation, of course, may be that sublethal toxicity rather than lethal toxicity plays some part in the elimination of species. Animals may be able to survive, for a short period of time, exposure to levels of metals similar to those found in the rivers, but may be unable to complete their entire life cycles under these conditions. In some cases, however, caddis larvae appear actually to become more abundant than expected in the polluted stretches of metalliferous rivers (Sprague *et al.*, 1965; Weatherley *et al.*, 1967; Norris *et al.*, 1982). Commonly, reduced predation pressure from fish has been suggested as a possible reason for this. There is also some inconsistency in the reported response of another important insect group, the Ephemeroptera (mayflies), to metal pollution. Some invertebrate groups which are commonly reduced or absent from polluted waters appear to be affected by alterations to the physical characteristics of the river bed which is caused by the particulate matter in mine effluents, rather than by the toxic action of the metals themselves. Oligochaetes, platyhelminthes and leeches are possible examples. Another form of stream-bed alteration which may have profound effects is the development of excessive algal growth, as reported in some metalliferous rivers in northern England (Armitage, 1980). One hypothesis to account for this process is that the metals, by removing herbivorous invertebrates, allow the establishment of an unusually luxurious algal mat on the river bed, which in turn interferes with the normal patterns of distribution of animals which are not themselves directly affected by the metal. Algal growth may, of course, be enhanced in rivers with substantial inputs of plant nutrients or organic matter, so this particular phenomenon may not be universal. However, there is evidence (see Chapter 4) that organic matter can afford some protection to aquatic species from the toxic effects of metals, so the outcome in any individual case may be difficult to predict. Indeed, no two rivers are exactly similar in their physical, chemical and biological characteristics, and the interactions of living organisms with one another and with their physical environment are so complex that whatever general pattern emerges, it will inevitably be subject to significant variations of detail according to the precise local circumstances. One major difficulty in studying the effects of pollution in the field and in drawing from them conclusions of general applicability is that it is usually impossible, in any individual case, to compare what is happening in the polluted river with what would have been happening had the river not been polluted. Most commonly, surveys of polluted rivers consist of contemporaneous observations of different sites along a river or within a river system at different distances from the source of pollution. However, it is well known that, even in the complete absence of pollution, the biological characteristics of different points within a river system vary widely according to the physical and chemical conditions prevailing at the different locations. The precise relationships between the biological community and its physical environment are not well understood, and it is therefore extremely difficult to distinguish the consequences of pollution from the response of the community to natural variations in its physical environment. In theory, it would be very interesting to have available for study two physically and chemically identical rivers, of which one was polluted (preferably with a single pollutant). In practice, such a pair of rivers probably does not exist, although attempts have been made on a limited scale to approach this situation experimentally. However, a pair of rivers which approaches this ideal as closely as we may reasonably hope is to be found in the North Pennine Orefield in northern England.

1.2 THE RIVERS EAST AND WEST ALLEN

The Rivers East and West Allen in Northumberland, England lie within the North Pennine Orefield. They flow roughly northwards for approximately 18 kilometres before joining to form the River Allen, itself a tributary of the South Tyne (Fig. 1.1). The rivers drain adjacent valleys which were heavily exploited for a variety of metal and mineral ores, particularly during the eighteenth and nineteenth centuries. After the early years of the present century, mining activity declined rapidly and virtually ceased in 1946. Sporadic mining for fluorspar took place at the Beaumont mine at the head of the East Allen until 1979, and a very limited amount of mining continued intermittently at Barneycraig on the West Allen until 1981, since when no activity has been recorded in either valley. Until the end of the nineteenth century, mining in the Allendales was dominated by the production of lead ore. The West Allen mines, in particular, produced substantial quantities of zinc ore, but this seems to have been simply discarded prior to 1899, being returned as backfill to disued shafts or left in

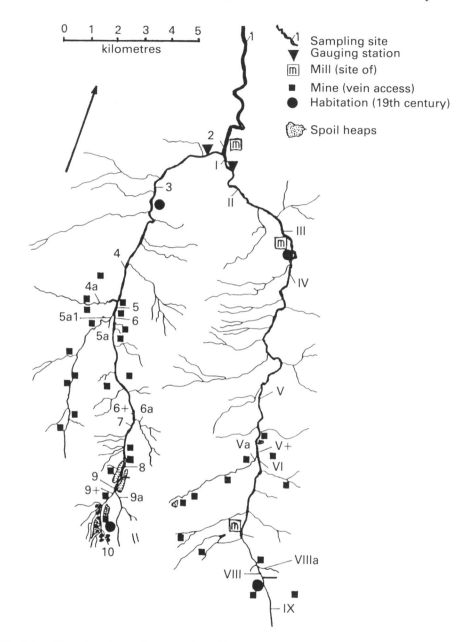

Fig. 1.1 — Map of the Rivers East and West Allen, showing sampling stations, derelict mines
and workings, and remaining spoil heaps. After Green (1984).

surface spoil heaps. Although the East Allen mines produced far greater quantities
of ore than those on the West Allen, the zinc-bearing veins are almost entirely
confined to the West Allen. Moreover, in the East Allen valley most of the waste

material was removed when the mines closed, while on the West Allen there remain extensive areas of spoil heaps and derelict land associated with mining activity. As a result, the River West Allen today contains considerably higher levels of zinc than the East Allen. A comparative study of these two physically-similar and geographically-adjacent rivers therefore provides a valuable opportunity to study the ecological effects of zinc.

1.2.1 Physical and chemical survey

In 1979 a series of 15 sampling stations were established on the West Allen and its tributaries. A further 12 stations on the East Allen were chosen for comparative study. The East Allen stations were chosen so that each was broadly similar in physical terms (width, depth, distance from source, substratum characteristics, nature of the surrounding terrain) to a site on the West Allen. Thus sites 7, 6a and 6+ on the West Allen could be considered as 'equivalent sites' to VI, Va and V+ on the East Allen (Fig. 1.1). Between June 1979 and October 1980, samples for chemical analysis were taken twice monthly from each site. The results are summarised in Fig. 1.2. In terms of temperature, pH, conductivity and a wide range of chemical

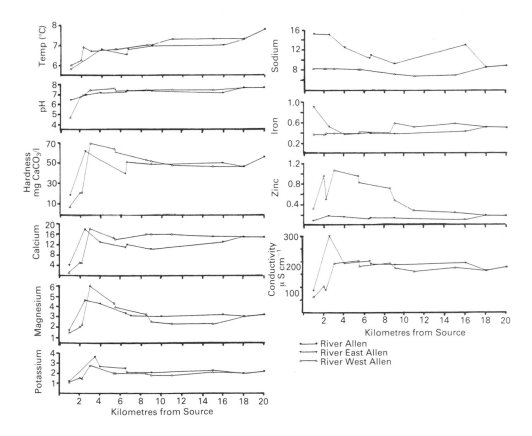

Fig. 1.2 — Median values of chemical determinands at mainstream sites on the Rivers East Allen, West Allen and Allen, plotted against distance from source (Green, 1984).

determinands the two rivers are very similar. Zinc levels in the West Allen were consistently up to ten times higher than in the East, placing some stretches of the West Allen among the most heavily zinc-polluted rivers and the East Allen at the upper end of the range found among rivers which may be considered unpolluted. Concentrations of toxic heavy metals other than zinc (copper, cadmium, lead, nickel, chromium) fell consistently below the detection limits of atomic absorption spectrophotometry. Some representative data for one pair of equivalent sites are given in Table 1.1, and levels of dissolved zinc for several pairs of sites are compared

Table 1.1 — Values of chemical variables for one pair of equivalent sites on the East and West Allen, 1979–80. Data from Abel & Green (1981). (Values as mg l^{-1} unless otherwise stated; ND = not detectable)

	Detection limit	West Allen Site 7		East Allen Site VI	
		Mean	Range	Mean	Range
pH		7.7	6.7–8.6	7.5	5.9–8.8
conductivity (μS cm^{-1})		257	61–460	214	62–330
Ca		33	6–73	38	5–65
Mg		6.39	1.35–13.8	3.9	1.57–13.8
K		2.5	0.5–4.1	2.9	0.8–5.2
Na		9.53	4.5–16	12.75	5.3–23.4
Mn		0.10	0.01–2.0	0.17	0.01–1.0
Fe		0.41	0.01–1.1	0.22	0.01–1.0
Pb	0.1		ND–trace		ND–trace
Cu	0.03	ND		ND	
Co	0.06	ND		ND	
Cd	0.01	ND		ND	
Ni	0.06	ND		ND	
Zn	0.001	1.31	0.45–3.68	0.13	<0.001–0.36

in Table 1.2. Compared to the soft, acidic Welsh rivers discussed earlier, the Allens are slightly alkaline and harder, though the zinc levels are broadly comparable.

1.2.2 Biological survey

Twenty-three sampling stations on the two river systems were surveyed in May, June and October of 1980. In addition certain sites were surveyed at monthly intervals during this period. Five replicate Surber samples (see section 3.3) were taken on each sampling occasion. The results of the survey indicated that the East Allen is, in biological terms, typical of many upland streams and rivers. The fauna is dominated by insects, with molluscs, crustacea, oligochaetes and platyhelminthes also repre-

Table 1.2 — Concentrations of filtrable zinc (mg l^{-1}) in river water from equivalent sites on the East and West Allen, 1979–80. Data from Abel & Green (1981). West Allen sites are designated by Arabic numerals, East Allen sites by Roman numerals

West Allen				East Allen			
Site	km from source	Zinc concentration		Site	km from source	Zinc concentration	
		Mean	Range			Mean	Range
10	1	0.24	0.04–0.56	IX	1	0.24	<0.001–0.83
8	3	1.88	0.65–4.15	VII	4	0.19	0.1–0.34
7	5.5	1.31	0.45–3.68	VI	6.5	0.13	<0.001–0.36
6a[a]	5.5	0.09	<0.001–0.33	Va[a]	6.5		0.04–0.49
6+	5.5	0.87	0.09–1.88	V+	6.5		0.11–0.14
5	9	0.51	0.12–1.22	V	9		<0.001–0.23
4	11	0.50	0.04–1.28	IV	13		0.08–0.49
3	15	0.34	0.08–0.80	III	17		0.04–0.31

[a] Denotes tributary site.

sented. In all 121 taxa were recorded from the Allens during the survey (this number has since increased considerably, partly through improved taxonomic techniques). As expected, the headwaters of the East Allen are relatively poor in species and numbers of individuals, but the fauna develops rapidly and in its lower reaches the East Allen is a moderately-good trout fishery. In contrast the West Allen, though initially similar to the East, receives zinc inputs about three kilometres from its source and the invertebrate fauna is markedly affected; some of the survey results are summarised in Fig. 1.3. Within the general pattern of reduced species diversity and numbers of individuals at the zinc-polluted sites, the effects of the zinc on particular invertebrate species could clearly be seen. Four species which were common in the East Allen were entirely absent from the mainstream of the West Allen: *Gammarus pulex* (Crustacea), *Ancylus fluviatilis* (Mollusca), *Taeniopteryx nebulosa* (Plecoptera), and *Hydroptila* sp. (Trichoptera). Several other species were absent from many stations on the West Allen, although present in the East (Table 1.3). Species found in both rivers were generally present in greatly reduced numbers in the West Allen; Fig. 1.4. shows some data for Plecoptera species, and similar patterns were shown by Ephemeroptera, Trichoptera, Coleoptera, and Diptera. Moreover, the survey data indicated that the life cycles of some species were altered in the zinc-polluted stretches. For example, nymphs of the stonefly *Amphinemoura sulcicollis* disappeared from the East Allen in July/August as the mature nymphs emerged from the water as adults. Nymphs of the new generation were abundantly re-established by October; in the West Allen, however, no new nymphs had appeared by October in the more polluted stretches, perhaps because the peak zinc concentrations tend to occur during the summer and have a particularly marked effect on the early life stages. Possibly the limited populations of this and similarly-affected species in the

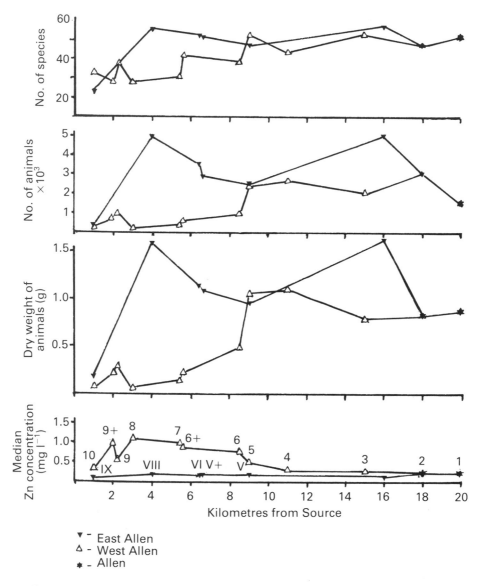

Fig. 1.3 — Total number of species, total number of animals and the total dry weight of animals collected from each sampling station during May — October 1980 in the Rivers East Allen, West Allen and Allen (Green, 1984).

West Allen are restored later in the year by downstream drift from unpolluted tributaries. The results of some further analyses of the survey data are described in section 3.4.

1.2.3   Zinc toxicity

The depletion of the invertebrate fauna in the zinc-polluted stretches may be due to the direct toxic action of the zinc, or may arise as a consequence of, for example, the

Table 1.3 — Species found along the length of the East Allen, but absent from threee or more sites on the mainstream of the West Allen. Data from Green (1984)

Plecoptera:	*Leuctra fusca*
	Siphonoperla torrentium
	Taeniopteryx nebulosa[a]
Ephemeroptera:	*Ecdyonurus venosus*
	Ecdyonurus dispar
	Baetis muticus
	Ephemerella ignita
	Caenis sp.
Trichoptera:	*Hydropsyche instabilis*
	Hydropsyche fulvipes
	Hydroptila sp[a]
Crustacea:	*Gammarus pulex*[a]
Mollusca:	*Ancylus fluviatilis*[a]

[a] Species absent from all sites on the mainstream West Allen.

disappearance of an animal's prey species or some other indirect cause. Green (1984) tested the lethal toxicity of zinc to several invertebrate species under conditions similar to those found in the Allens. One way to measure lethal toxicity is to detemine the concentration of the poison which will kill half of a sample of animals within a specific period, such as four days — this value is termed the 96 hour median lethal concentration, or 96 h LC50. Methods of measuring toxicity are discussed in detail in Chapter 4. Table 1.4 shows some results. All of the species tested were either absent from all or part of the West Allen mainstream, or found only in greatly reduced numbers. Comparing the zinc levels in the water (Table 1.2) with the values in Table 1.4, three groups of animals can be distinguished. First, those such as *Gammarus pulex* and the snail *Lymnaea peregra*, for which the LC50 values are very close to the zinc concentrations found in the river. It is reasonable to conclude that the river water is rapidly toxic to these species, and that any individuals which found themselves in the polluted stretches would quickly die if they were unable to escape speedily to a less polluted area. Secondly, there is a group of species which appears to be remarkably resistant to zinc. The stonefly *Chloroperla tripunctata*, a caddis fly of the genus *Limnephilus*, and the beetle *Deronectes depressus* withstood concentrations of zinc greater than 360 mg. l^{-1} for up to 12 days without any animals dying at all. This suggests that although the possibility of sublethal toxicity cannot be discounted, it is likely that mechanisms other than the direct effect of zinc on the animals are largely involved. Thirdly, there is an intermediate group for which median lethal zinc concentrations are roughly between 10 and 100 times higher than the zinc levels recorded in the river. For this group, indirect effects may also be involved; however, experience with many poisons and animals, particularly fish, which have been more extensively studied suggests that sublethal toxicity of zinc is likely to be an important factor in the observed ecological status of these species.

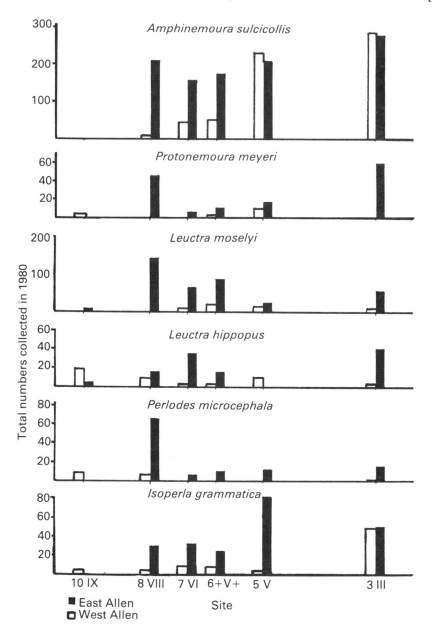

Fig. 1.4 — Total numbers of selected species of Plecoptera (stoneflies) from 'equivalent sites' on the Rivers East and West Allen during May–October 1980 (Green, 1984).

Some species were tested on more than one occasion at different times of the year and the median lethal concentrations varied considerably (Table 1.4). In part, this is probably a reflection of the fact that toxicity often cannot be measured, however it is

Table 1.4 — Toxicity of zinc to some invertebrate species. Data extracted from Green (1984). Where two sets of results are given, the results represent the range of toxicity values recorded in replicate tests carried out at different times

Species	Result[a]	
Gammarus pulex	2.0(1.67–2.40)	336 h LC50 0.66 (0.50–0.88)
Baetis rhodani	1.3(0.6–2.9)	
	31(16.3–58.9)	336 h LC50 16.5 (7.5–36.3)
Rhithrogena semicolorata	70(36.8–133)	336 h LC50 52 (34.4–78.5)
	135(71–256)	336 h LC50 68 (45.3–102)
Leuctra moselyi	15.7(9.24–26.7)	
	55(30.6–99)	336 h LC50 17.5 (4.07–75.3)
Amphinemoura sulcicollis	120 h LC50 130 (59.1–286)	
Isoperla grammatica	90(40.9–198)	
Lymnaea peregra	2.6(0.8–8.3)	
Chloroperla tripunctata	No mortalities in 144 h at concentrations >360 mg l^{-1}	
Deronectes depressus	No mortalities in 14 h at concentrations >360 mg l^{-1}	
Limnephilus sp.	No mortalities in 288 h at concentrations >360 mg l^{-1}	

[a] 96 h LC50 values in mg l^{-1} unless otherwise stated.
Values in parentheses are 95% confidence limits.

expressed, with great precision. However, it is possible that some species vary in their susceptibility to zinc at different times of the year, according to the stage of the life cycle which they have reached. The effect seems to be particularly marked in the case of the mayfly *Baetis rhodani*. This feature of the results, along with the various aspects of sublethal toxicity and the nature and extent of the indirect effects of zinc, all need to be further investigated before the effects of the zinc on the receiving water fauna can be fully understood.

1.2.4 Some unresolved questions
The investigations of the Allens and other metalliferous rivers have shown that in each case, the effects of the metal on the receiving water fauna are broadly similar. Generally, it is observed that there is a reduction in the number of species present in the polluted areas, together with a reduction in numbers of individuals of those species which are not eliminated altogether. Certain species appear to be particularly sensitive to heavy metals, others apparently more resistant, and the status of some species with respect to metal pollution appears to vary from one location to another. In other words, there is a perceptible general pattern which is subject to variations in detail. It is important to know to what extent the findings of any particular study are of general application, and to what extent they are restricted to, for example, particular geographic regions, or to waters which have particular physical, chemical or biological characteristics. Therefore it is necessary to consider in more detail the

ecotoxicological mechanisms which underlie the observed ecological effects of the metals. Some of these are currently under investigation, in relation to the Allens and other rivers elsewhere. However, the purpose of this section is not to describe these investigations in detail, but to illustrate some of the biological questions which may arise out of the investigation of a polluted water.

The role of sublethal toxicity is potentially an important area for further investigation. The concentrations of zinc which are lethal to several of the affected species lie between 10 and 100 times higher than the zinc levels in the water, that is, within the range where sublethal toxicity may be expected. In section 1.3 it is shown that for fish, if the concentration of poison present in the water exceeds a small fraction of the lethal concentration for only a small percentage of the time, fish will be absent from rivers. Sublethal toxic effects in invertebrates have been less extensively studied than in fishes, but some examples arising from the investigation of the Allens have been investigated. In laboratory experiments with detritivorous caddis larvae of the genus *Limnephilus*, the feeding rates of animals exposed to zinc at the concentrations found in the West Allen were 30% lower than the controls (Abel & Green, 1981). Although this species appeared highly resistant to the lethal action of zinc, possibly its inability to feed, and therefore grow, normally in the presence of zinc accounts for its absence. Many aquatic animals have been shown to be able to detect and avoid pollutants in laboratory choice chambers. For invertebrates in fast-flowing rivers this effect is likely to be manifested through an increased rate of downstream drift, leading to depopulation of the polluted stretches. Gilhooley (1988) found that the rate of drift of *Gammarus pulex* in an artificial river channel was indeed increased in the presence of zinc.

Another aspect of sublethal toxicity is the possible accumulation of zinc in the animal tissues. Animals may survive exposure to low levels of metal without apparent effect, but may continue to accumulate metal from solution, by ingestion of or contact with particulates or from their food, until harmful levels of metal in the tissues are reached. Similar considerations may apply in the case of organic pollutants which are not readily metabolised, such as certain pesticides.

The effects of other poisons, and of environmental factors modifying toxicity, should be considered. In many of the rivers referred to above, more than one metal was present in significant quantities. In the zinc-polluted River Nent in the North Pennine Orefield, substantial inputs of organic farm waste were recorded (Armitage, 1980). Clearly, some of the differences observed between different rivers may be attributable to such confounding factors. In the Allens, no significant source of pollution apart from mine effluents and runoff from waste heaps has been located. Zinc is overwhelmingly predominant among the heavy metals (Green, 1984). Cadmium is often associated with zinc in mining wastes, and is potentially highly toxic. However, even in the undiluted mine effluents cadmium rarely reaches a concentration of 0.01 mg.l^{-1} and the Allens appear to be unusually low in cadmium. While it is therefore tempting to ascribe the disturbance of the West Allen solely to the presence of zinc (particularly in the case of molluscs and crustaceans, where the toxicological evidence readily supports this view), some caution must still be exercised. In the river Wye, which is not polluted, Edwards *et al.* (1978) found that *Gammarus pulex* was absent from sites where the calcium concentration in the water was below 10 mg.l^{-1}. In the West Allen the calcium concentration is normally above this level, but does occasionally fall below it. Thus low calcium may be at least partly responsible for the absence of this species. Zinc is, of course, also generally more toxic in soft water anyway (see section 4.2).

The partitioning of the zinc in the river environment is likely to have a great bearing on the correct interpretation of toxicological data. The precise form in which the zinc is present can have a large influence on its toxicity (see section 4.2). Metals may exist, in the aquatic environment, in dissolved, colloidal or particulate form; in two or more different oxidation states; as simple ions, inorganic complexes or organometal complexes. The biological and toxic properties of these different forms may vary greatly. In laboratory tests, the metal is normally presented to the animals in dissolved form, but it is almost certainly never entirely in the form of simple ions. In the river, the measurements made of 'dissolved' metals are, more accurately, of that fraction of the metals which can pass through a 0.45 μm filter, which is not necessarily the same thing. Another aspect of partitioning is that many animals burrow into the substratum, and may be exposed to levels of zinc in the substratum which are much higher than those in the overlying water. Green (1984) devised an apparatus for sampling the interstitial water without contamination with overlying water, and in the Allens found zinc levels in the interstitial water up to 25 times higher than those in the water flowing over the substratum. All of these factors complicate the interpretation of toxicological data and their application to the situation in the field.

Some of the observed effects of the zinc in the river may be due to the ecological consequences of selective toxicity. Zinc, like most pollutants, affects some species more than others. Species which are not directly affected may nevertheless respond to the selective removal of more sensitive species. For example, if a predator is deprived by selective toxicity of its prey, it may itself be reduced in numbers, or be eliminated. Alternatively, it may be able to find a different prey species, which may then also show a numerical response. Where two competing species are unequally affected by the pollutant, both may show a change in their abundance or pattern of distribution. Thus the consequences of pollution can never be fully understood without some knowledge of trophic, competitive and other interspecific relationships.

Finally, the effects of the pollutant on the decomposer organisms of the river may be considered. The productivity of many aquatic ecosystems, particularly rivers, is sustained by the input of allochthonous organic matter such as dead leaves. Decomposer organisms — bacteria and fungi especially — are thus of great importance in making available to the river fauna the major energy source represented by detritus. Indeed it is likely that most aquatic animals, like animals generally, cannot readily digest plant material unaided by micro-organisms. An invertebrate may appear to be eating a dead leaf, but in reality is probably obtaining its nutrition from the bacteria, fungi, protozoa and microinvertebrates which have colonised the leaf and processed it into a form which is more accessible to detritivorous macroinvertebrates. Zinc is known as a fairly potent bactericide and fungicide. Possibly, therefore, one effect of the zinc is to interfere with the processing of detritus by decomposer organisms, thus depriving the invertebrates of a major food source and being indirectly responsible for the depletion of the fauna.

These and similar questions will be recurring themes in the following chapters. The Allens represent a fairly simple and straightforward case of water pollution, but their study has given rise to questions which can as yet scarcely be answered. The next case study, however, shows that with a suitable combination of field and laboratory studies, detailed investigation of the effects of pollutants can provide an empirical basis for tackling practical problems relating to the control of pollution and the management of polluted waters.

1.3   TOXICITY AND THE STATUS OF FISHERIES

Our second case study illustrates the way in which toxicological data obtained from laboratory investigations can be combined with chemical and biological data from field surveys into a predictive model which suggests specific measures to protect or improve the status of fisheries in polluted waters. The example is particularly instructive because even today, toxicological research is still dominated by the measurement of lethal toxicity. There is, therefore, a widely-held view that a great deal of toxicological research is misdirected, irrelevant or of limited value in the actual management of polluted waters, since the most intractable problems in practice arise from sublethal, rather than lethal, toxic effects. The following example shows that this view is based on an inadequate understanding of the applications of toxicological data. The account is based upon the application by Alabaster *et al.* (1972) of the results of a long series of investigations carried out by what was then the Water Pollution Research Laboratory, a Government establishment in Stevenage, UK. These investigations were based upon the work of a substantial number of people carried out over a period of nearly 20 years. The results can be synthesised into an empirical relationship between the presence of certain common toxic pollutants and the ecological status of aquatic communities which can be used for specific management purposes. For example, when several pollutants are present, it allows the identification of those which are responsible for the greatest adverse effects, so that pollution control measures can be selectively directed towards those pollutants whose removal would lead to the greatest improvement. It also allows predictions to be made of the likely effects of additional pollution, or of physical changes in the receiving water environment.

In the more heavily polluted rivers of Britain and of similar industrialised countries, the most abundant toxic pollutants are copper, zinc, phenol, cyanides, and ammonia. The toxicity of these poisons to the rainbow trout, *Salmo gairdneri*, was first studied in great detail. The rainbow trout was chosen because it is a widely-available species, is amenable to life in the laboratory, and is a fish of considerable commercial importance. Consequently a great deal is known about many aspects of its biology. Additionally, it is sensitive to most toxic pollutants and reacts more quickly than most species to adverse environmental conditions. Within a few years, the lethal toxicity of the common pollutants to this species was reliably determined. The effects of common environmental variables on pollutant toxicity was also studied. In particular temperature, water hardness, pH and dissolved oxygen concentration were found to have significant effects on the toxicity of many pollutants. Finally, methods were devised for the study of the effects of fluctuating concentrations of poisons, and for determining the toxicity of poisons in different combinations of varying composition. Much of this work is described in greater detail in Chapter 4.

A crucial idea in the development of this approach is that of the unit of toxicity, or toxic unit. A toxic unit was defined as the concentration of a pollutant which would kill half of a sample of rainbow trout in 48 hours (i.e. the 48 hour median lethal concentration or 48 h LC50 as described in Chapter 4). Because trout react quickly to most poisons at lethal concentrations, the 48 h LC50 is close to the lethal threshold concentration, that is, the concentration which would just kill half the sample of fish during an exposure of indefinite duration. A few simple examples will illustrate how toxic units are used. Assume that, under a certain set of conditions, it is found that the concentration of zinc which will kill half the trout in 48 hours is 2 mg l^{-1}. Under these conditions, one toxic unit of zinc is 2 mg l^{-1}. Now, assume that the experiment is repeated under similar conditions, but in water which has a greater degree of hardness. It may be found that the 48 h LC50 for zinc is now 10 mg l^{-1}. Under these new conditions, one toxic unit of zinc is 10 mg l^{-1}, and 2 mg l^{-1} of zinc is now equivalent to only 0.2 toxic units. Now assume that we wish to measure the toxicity of copper to trout. In soft water, one toxic unit of copper may be equal to 0.5 mg l^{-1} and in hard water 2.5 mg l^{-1}. Now assume that we wish to know whether, in the soft water, fish would survive in the presence of 1.5 mg l^{-1} of zinc together with 0.25 mg l^{-1} copper. In this case, 1.5 m l^{-1} zinc equals 0.75 toxic units, and 0.25 mg l^{-1} copper equals 0.5 toxic units. The total number of toxic units is therefore 1.25, whereas we know that by definition, 1 toxic unit will kill half the fish in 48 hours. Therefore we would expect that significantly more than half of the fish would die within 48 hours. In the hard water, however, zinc would contribute 0.15 toxic units, and copper 0.1 toxic units, a total of only 0.25 toxic units. We might therefore expect that the majority of fish would survive these concentrations of zinc and copper in the hard water. These expectations assume, of course, that the effect of the poisons in combination is neither more nor less than the sum of their individual effects. This point has been extensively investigated (see section 4.2.3) and fortunately, with some important exceptions, appears to be generally true for this group of pollutants. It is not true, however, for all combinations of pollutants. The toxicity of mixtures of poisons is discussed more fully in Chapter 4.

The next stage is to use this technique to estimate the expected toxicity of polluted river water to fish, and to compare the results with the observed status of fish populations. It is not difficult to measure the concentrations of zinc, copper, phenol, cyanides and ammonia in samples of river water. The temperature, pH, hardness and dissolved oxygen concentration of the water at the time of sampling must also be determined. Using these data, the measured concentrations of each pollutant, in mg l^{-1}, can be converted into toxic units. Summing these 'fractional toxicities' will give a measure, in toxic units, of the total toxicity of the river water. Obviously, the toxicity of the river water will vary with time, depending upon the amount of effluent being discharged, the quantity of water available for dilution, and the prevailing environmental factors. It would not be sensible to try to associate the observed status of the fish population of a stretch of river with the toxicity of the water on any single occasion. It is necessary to take a relatively large number of samples over a reasonably long period of time, ideally a whole year, and to calculate the toxicity of the water on each occasion. Then a graph of the kind shown in Fig. 1.5 is constructed. Each sampling station investigated generates one line on the graph.

The line on the left of Fig. 1.5 represents the distribution of toxicity over time at a

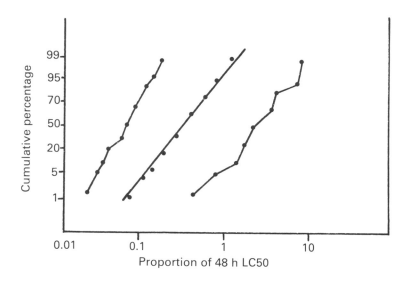

Fig. 1.5 — Distributions of toxicity calculated by the method of Alabaster *et al.* (1972). The line on the left is typical of a station which supports fish; the line on the right represents a fishless station; and the central line is reconstructed from the boundary distribution of toxicity between fishless and fish-supporting stations in the River Trent catchment. See text for explanation.

sampling station which sustains a fish population. It is constructed by plotting the cumulative frequency with which a given level of toxicity occurs in a number of samples taken from the site under investigations over a period of time. It is convenient to plot the graph using a logarithmic scale for the toxicity values and a probability scale for the *y*-axis. Thus, in the example shown, on 5% of sampling occasions the toxicity was equal to or less than 0.04 toxic units. On 20% of occasions, it was equal to or less than 0.055 toxic units, on 50% of occasions it was equal to or less than 0.08 toxic units, and so on. The line on the right of Fig. 1.5 represents a toxicity distribution typical of a fishless station. The increased toxicity of the water is represented by a shift of the line towards the right. Plotting a large number of these lines (one for each of the sampling stations studied) showed that the lines were plotted in various places on the diagram, depending upon the distributions of toxicity recorded at each station. However, lines corresponding to stations where fish were normally present were grouped to the left of the diagram, and lines corresponding to stations which were normally fishless were grouped to the right; a line can be contructed through the narrow zone of demarcation between the two groups. From this line it is possible to read off the 'co-ordinates of boundary distribution between fishless and fish-supporting waters' (Alabaster *et al.*, 1972). The central line in Fig. 1.5 has been reconstructed from the co-ordinates determined by Alabaster *et al.* (1972) for sampling stations in the catchment area of the River Trent. The values of the co-ordinates are shown in Table 1.5.

These data show that a water will only sustain a fish population if for at least 50% of the time the toxicity of the water is less than 0.28 toxic units; and if for 90% of the time its toxicity is less than 0.6 toxic units, for 95% of the time less than 0.73 toxic

Table 1.5 — Co-ordinates of approximate boundary distribution of 48 h LC50 between fishless and fish-supporting waters in the River Trent catchment area, England. See Fig. 1.5 and text for explanation. Data from Alabaster *et al.* (1972)

Per cent probability	1	5	10	25	50	75	90	95	99
Sum of proportions of LC50	0.07	0.10	0.13	0.18	0.28	0.41	0.60	0.73	1.1

units, and so on. Where the toxicity of the water exceeds 1.0 toxic unit for as little as 2 or 3% of the time, fish will generally be absent. The simplest way to determine whether or not a sampling station is likely to support fish is to draw the line of toxicity distribution. If the line, or any substantial portion of the line, falls to the right of the boundary line, the water will be unlikely to support fish. The further to the left the line falls, the smaller is the likely effect of toxic pollution on the fish population. Where the line falls to the left of the boundary but close to it, the fishery may be of only marginal quality.

Although this relationship between toxicity and fishery status is purely empirical, it has considerable potential value in water management, as the following examples show. Assume that a stretch of river is fishless, and that we wish to know the most effective means of re-establishing a fish population. Chemical analysis of the water on a number of occasions reveals that several pollutants are present, and that the dissolved oxygen concentrations are generally low. Many poisons are more toxic at low dissolved oxygen levels. Could the fishery be restored by increasing dissolved oxygen levels, or is the removal of specific toxic substances required? In the toxicity distribution represented by the line on the right of Fig. 1.5, the toxicity has been calculated using the dissolved oxygen calculations measured on each sampling occasion. These values can be recalculated, on the assumption that the dissolved oxygen concentration will be increased to any desired level. A new line can therefore be constructed using the hypothetical oxygen values, and this line will be displaced to the left. If the displacement carries the line beyond the boundary distribution, it is expected that a fishery could be re-established by increasing the levels of dissolved oxygen in the water. This could be achieved relatively easily and cheaply, for example by imposing stricter controls on the discharge of sewage or organic wastes which cause reduced dissolved oxygen levels; by upgrading sewage treatment plants; or by utilising one of the various methods available for aerating rivers such as constructing a weir. If, however, the displacement does not carry the line beyond the boundary distribution, the desired effect is only likely to be achieved by the removal of specific toxic substances, which may be more difficult and expensive. As similar line of reasoning can be applied in situations where, for example, it is proposed to site a new discharge on a river and we wish to predict the likely effect of the new discharge. The new discharge may contain additional toxic pollutants, and/or may alter the temperature, dissolved oxygen or other environmental characteristics of the receiving water. If the effect of the proposed new conditions is significantly to shift the existing distribution of toxicity towards the boundary conditions, or to carry the existing distribution beyond the boundary conditions, the decision may well be made that the proposed new discharge is unacceptable, or should be subjected to more rigorous control.

A further application of this technique is the determination of the relative contributions of each of several pollutants to the overall toxicity of the water. It is frequently the case that several pollutants are present in significant quantities, but chemical analysis alone cannot reveal which of them is exerting the most serious adverse effect on the biota. If the most significant pollutants can be identified, specific control measures may be directed against those, rather than against pollutants whose biological impact may be small, thus affording greater efficiency in the allocation of resources. Alternatively, we may wish to know whether the addition of a new pollutant, or an increase in the expected concentration of an existing pollutant, is likely to have any serious effect. Since the fractional toxicities of each pollutant are initially determined separately, plotting them in the manner shown in Fig. 1.6 allows

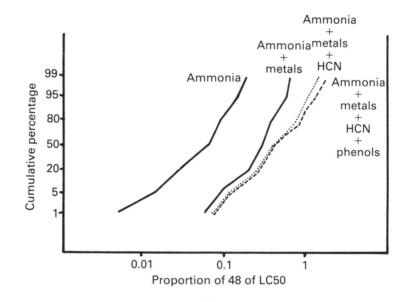

Fig. 1.6 — Contribution to toxicity from ammonia, metals, hydrogen cyanide and phenols at a heavy-polluted site in the River Trent catchment (Alabaster *et al.*, 1972).

these questions to be answered. In one example given by Alabaster *et al.* (1972), the toxicity distribution of ammonia alone was plotted, giving a line similar to that shown on the left of Fig. 1.6. A second line, showing the toxicity distribution of metals and ammonia combined, displaced the line significantly to the right. This indicates that metals were contributing substantially to the overall toxicty of the water. A third line, showing the effect of including cyanide in the calculation, similarly indicated a smaller, but substantial contribution from cyanide. Phenol, however, contributed little to the overall toxicity, although it was present in the water; its inclusion in the calculation gave only a very slight displacement to the right. Using the same method, the authors were able to show that among the metals present, copper and zinc were identifiable as major pollutants in the sites they studied, whereas nickel, chromium and cadmium contributed little to the overall toxicity of the water. This was in spite of the fact that nickel alone formed 20% of the total metal present in chemical analyses. However, nickel and, in particular, cadmium are known to be very slow-acting poisons (see Chapter 4). The model in use relies upon the 48 h LC50 as the definition of the toxic unit, and this is only satisfactory if the 48 h LC50 is a good approximation of the lethal threshold concentration. Since this is not the case for nickel and cadmium, the calculations were revised using the lethal threshold concentration as the definition of the toxic unit. Under these circumstances, the results indicated that nickel and cadmium (but not, in the cases studied, chromium) were significant pollutants in some instances.

The synthesis of data from field and laboratory studies into the potentially useful model for the management of polluted waters has many advantages. Toxicological data, particularly those derived from the study of lethal toxicity, can rapidly be accumulated and replicated under controlled conditions, but are difficult to apply to real situations in the field because our knowledge of ecotoxicological mechanisms is inadequate. We might expect, for example, that if the level of pollution in the water was equal to the 48 h LC50, that half or more of the fish would die and that sublethal toxicity and mechanisms involving the relationships of fish with other members of the aquatic community might be sufficient to account for the total absence of fish from a particular site. The empirical finding that if the toxicity exceeds 1.1 toxic unit for 1% of the time, or exceeds 0.07 toxic units for 99% of the time (Table 1.5) is, however, more directly useful to us. It does not actually matter, in the day-to-day management of the water, why exactly this happens — whether, for example, it arises through sublethal toxicity, avoidance by the fish of specific adverse conditions which they have detected, or because the pollutants are adversely affecting the food organisms of the fish. Knowledge of ecotoxicological mechanisms is ultimately valuable because it allows the interpretation and application of data in the frequent circumstances that decisions have to be made on the basis of incomplete information. In these same circumstances, purely empirical relationships are also useful for precisely the same reasons. An alternative approach would be to rely entirely upon data from field surveys. In principle, any particular management decision could be formulated with reference to experience of what has happened previously in a similar situation elsewhere. In practice, the data base of previous information which would be required is so vast that even if it existed, the task of collating it and analysing it in such a way as to extract from it the specific information required would be impossible. In any case, new situations are constantly arising; and water bodies differ so widely in their physical, chemical and biological characteristics that purely anecdotal evidence is of limited predictive value without some attempt to understand and apply fundamental principles. In practice, the control and amelioration of the problems raised by water pollution are best achieved by a combination of approaches — field-based and laboratory-based, empirical and fundamental.

2

Sources and effects of water pollutants

There are hundreds, perhaps thousands of pollutants whose effects are of actual or potential concern. Their numbers increase annually, as new compounds and formulations are synthesised. A substantial minority of these find commercial applications and become significant pollutants of water during their manufacture and in subsequent use. It is clearly impossible within the scope of a short book to discuss all of these in detail; instead, the sources and effects of broad categories of pollutants will be discussed in general terms, and where appropriate reference will be made to further sources of information.

2.1 THE NATURE OF EFFLUENTS

Water pollution is most commonly associated with the discharge of effluents from sewers or sewage treatment plants, drains, and factories. Outfalls of this kind are known as 'point-source discharges'. Most cases of accidental, negligent or illegal discharge are also from point sources. The concentration of pollutant in the receiving water is initially high, decreasing as the distance from the point of discharge increases. The effects of the pollution are therefore frequently easy to observe. Some of the more serious forms of pollution arise, however, from 'diffuse' sources, that is the pollutant does not enter the water from a single point. For example, in agricultural areas, surface water runoff and groundwater infiltration into lakes and rivers can introduce plant nutrients (from fertilisers) and pesticides in substantial quantities to water bodies. The effects of pollution from diffuse sources can be serious, but are often less immediately obvious than those from point sources as there is no adjacent unpolluted area with which comparisons may be made. Many pollutants also enter water through fallout from the atmosphere.

Most effluents are complex mixtures of a large number of different harmful agents. These include toxic substances of many kinds, extreme levels of suspended solids, and dissolved and particulate putrescible organic matter. In addition, many effluents are hot, of extreme pH value, and normally contain high levels of dissolved salts. Detailed compilations of data on the composition of sewage and industrial effluents of many kinds are given by Bond & Straub (1974), and by Sittig (1975). Some representative values for treated sewage effluent are given in Tables 2.1 and 2.2.

Most effluents also vary in their strength and composition, on a seasonal, diurnal or even hourly basis. Most sewage treatment plants report regular diurnal peaks and troughs in their output according to patterns of water use. Sometimes storm-water drains are connected to the sewerage system, so the strength of the sewage effluent will vary with rainfall. Alterations in the strength and composition of sewage also influence the efficiency of the sewage treatment process, so that dilution of the influent does not necessarily cause an improvement in the quality of the effluent. In industrial plants, variations in the quality of the raw materials, or changes in specification of the finished product, will require changes in the operating conditions of the plant and lead to changes in the composition of the effluent. Many industrial processes are 'batch' rather than continuous processes, so that some effluent discharges will be intermittent rather than continuous.

Nevertheless, it is often possible to generalise about the effects of different kinds of effluent on their receiving waters. Broadly speaking, the effects of sewage effluent, kraft pulp mill effluent, coal mine effluent and so on are consistent wherever they occur. However, a detailed understanding of the effects of individual components of an effluent, and of the consequences of variation in the composition of an effluent, is essential for pollution control. Wastewater treatment is expensive, and in order to devise cost-effective treatment processes it is necessary to identify those components of the effluent which cause the greatest damage to the environment. This is because it is usually impossible to devise an economically-feasible process which will be equally effective against all the components of a complex effluent. Frequently, treatment of an effluent to remove one component will exacerbate the problem of removing another; it is notoriously difficult, for example, to treat satisfactorily effluents which contain both cyanides and phenolic compounds, although many basic industrial processes produce just such an effluent. Therefore in order to devise the optimum pollution control strategy it is necessary to study, in the laboratory and in the field, the effects of effluents and of their individual components.

2.2   THE ENVIRONMENTAL REQUIREMENTS OF AQUATIC ORGANISMS

The effects of pollutants in aquatic ecosystems cannot be understood without some knowledge of the ecophysiology and basic biology of the aquatic biota. Those aspects which have most bearing on the interaction between water pollutants and aquatic organisms are briefly discussed here. There are many authoritative and readable texts on freshwater ecology which may be recommended to readers who require a fuller treatment. Among these are Maitland (1978), Moss (1980), Hynes (1970) and Whitton (1975).

In order to survive, a living organism must spend its life in an environment which meets its needs: a suitable physical habitat which provides space, shelter, and a sufficient supply of food, oxygen, and other metabolic requirements; and which is not subject to extremes of temperature or other physical variables which lie outside the range which the organism can tolerate. Obviously, different habitats have very different physical characteristics, and organisms have evolved a fascinating array of adaptations which have enabled them to colonise every part of the Earth's surface.

Table 2.1 — Analysis of sewage effluents after primary and secondary treatment. Figures given are in mg l^{-1}, and are the range of values found at four different treatment works, as summarised by Bond & Straub (1974). Treatment plants operating under adverse conditions yield higher values than those shown here

Analysis	Range
Total solids	640–1167
Suspended solids	15–51
Biological oxygen demand	2–70
Chemical oxygen demand	31–155
Organic carbon	13–20
Anionic detergents	0.75–1.4
Ammonia	1.9–22
Nitrate	0.25–38
Nitrite	0.2–1.8
Chloride	69–300
Sulphate	61–270
Phosphate	6.2–9.6
Sodium	144–243
Potassium	20–26

Table 2.2 — Levels of heavy metals recorded in the final effluent of a typical sewage treatment plant operating under good conditions. Data from Bond & Straub (1974)

Metal	Concentration range ($\mu g\, l^{-1}$)	Typical % removal compared to raw sewage
Zn	85–190	65
Al	460–550	77.5
Fe	160–290	70.5
Cr	36–70	76
Cu	31–38	60.5
Pb	‹20	
Ag	11–12	17.3
Cd	‹20	
Ni	‹10	
Sr	280–450	17.6
B	240–260	13.3

To a greater or lesser extent, every living organism is adapted — in its morphology, physiology and behaviour — to the environment it normally inhabits. Some are remarkably specialised, that is they are adapted to specific places and/or modes of life in which they are very successful, but are excluded from living in most habitats. Others are more generalised in their adaptations, perhaps being nowhere particularly abundant but able to survive reasonably well in a wider range of habitats. Few organisms, however, are universally distributed.

Any of a living organism's many individual requirements may act as a *limiting factor* preventing the establishment, survival or reproduction of a species in a particular habitat. Aquatic plants for example, are commonly limited in their distribution and abundance by the availability of nutrients, such as phosphorus or nitrogen. An abundance of phosphorus is of no use to the plant if it has no nitrogen, and vice versa. Further, the nutrients must be present in a form which the plant can use. Photosynthetic plants, of course, also require light, an important limiting factor in most aquatic habitats. The non-photosynthetic flora (fungi, bacteria) are more likely to be influenced by the levels of dissolved or particulate organic material present. Animals, in turn, are greatly influenced by the quality and quantity of the aquatic flora, because many animals rely upon plants for food, shelter, as a repository for eggs, and so on. Animals are also influenced by the physical environment — current speed, nature of the substratum, temperature — but in addition are generally much more susceptible than plants and bacteria to the prevailing levels of oxygen.

Although many aquatic animals are air-breathers, the majority have to obtain their oxygen from water. Oxygen is not very soluble in water, and water is a rather heavy and viscous medium. Moving through water requires a great deal of energy expenditure, and therefore a high oxygen consumption. To obtain from the water the meagre amount of oxygen dissolved therein requires that the respiratory surfaces be moved through the water, or the water be moved over the respiratory surfaces. Further, as the water temperature increases, the solubility of oxygen in water decreases, while the oxygen requirement of the animal (fish or invertebrate) actually increases. Thus the survival of animals in water is crucially dependent upon the extent to which their oxygen demand can be matched to the availability of oxygen from the environment. Alexander (1970) describes elegantly and in detail the respiratory predicament of fishes. A fish at rest in well-oxygenated water ventilates its gills fairly slowly. Respiratory exchange is efficient, and the fish removes most of the oxygen from the water passing over its gills. If, however, the fish ventilates more quickly, as it must for example if it becomes physically active, the water passing over the gills has less time to equilibrate with the blood, and the efficiency of respiratory exchange drops. Thus, in order to double the rate of oxygen uptake, the fish must more than double the amount of water pumped over the gills. Measurements of the ventilatory mechanisms in fish suggest that a resting fish can remove 80% of the oxygen in the water passing over the gills, but this figure drops to about 30% in an actively swimming fish. To increase its oxygen uptake by a factor of five, a fish must pump about fifteen times as much water per unit of time. The weight and viscosity of water are such that the muscles which pump the water over the gills themselves consume a significant proportion of the oxygen obtained from the water. This proportion increases dramatically as the fish's activity increases, or as the temperature rises, or as the dissolved oxygen level in the water falls. Alexander (1970) calculated that a resting fish in water containing 8 ml O_2 per litre might under typical conditions expend 0.025 ml of oxygen to work the respiratory muscles, in order to obtain 6 ml of oxygen. However, if the oxygen concentration in the water fell to 1 ml per litre, the fish would expend 0.25 ml of oxygen in order to obtain only 0.3 ml. At this point, the fish is clearly close to asphyxiation and has no ability to engage in physical activity.

Clearly, oxygen is an important limiting factor to fishes. Equally clearly, however, limiting factors interact with one another. An oxygen concentration which is acceptable at 10°C is limiting, even lethal, at 20°C. Varley (1967) described very clearly the interactions of temperature and oxygen and their influence on the distribution patterns of freshwater fishes in Britain. Similar principles are involved among the invertebrates. Nymphs of stoneflies (Plecoptera), for example, are particularly fastidious in their requirements for oxygen, and are absent from waters whose oxygen concentration drops much below saturation value for appreciable periods of time. Ephemeroptera (mayfly) nymphs are a little more tolerant, but are generally more sensitive to oxygen depletion than Trichoptera (caddis fly) larvae. The amphipod crustacean *Gammarus pulex* and the isopod *Asellus aquaticus* are remarkably similar in their habits and environmental requirements, but above a critical oxygen concentration *Gammarus* predominates, while below that level it tends to be replaced by *Asellus*. Within major taxonomic groupings, the distribution of individual species is correlated with dissolved oxygen levels in the water. Indeed, the response of invertebrates to the prevailing levels of dissolved oxygen is so well-known that their distribution patterns can be used as indicators of the prevailing environmental conditions (see Chapter 3).

The patterns of distribution and abundance of organisms cannot, however, be explained solely in ecophysiological terms. A habitat may be perfectly adequate, physically and chemically, for a herbivorous animal, but if the habitat is not suitable for the animal's food plants the herbivore will be absent. Where a habitat is marginally acceptable to a species, the organism may nevertheless be absent because another species, with similar requirements, is better adapted to those conditions and competes more successfully for the available resources. Some organisms are absent from particular habitats not because the habitats are unsuitable, but simply because the species concerned have so far failed to overcome a geographical barrier which prevents their spread. In other words, the observed patterns of distribution and abundance of species are the net result of a complex interaction of physical, chemical and biological factors. Nevertheless, within a habitat it is frequently possible to observe a zonation pattern in the distribution of species in response to directional changes in the physical environment. The zones may be sharply delineated, as on a steeply-sloping sea shore, or they may merge gradually one into another, as is more typically found along the length of a river (Hynes, 1970; Hawkes, 1975). Pollution may directly influence one or more components of the dynamic equilibrium of physical, chemical and biological phenomena which gives an ecological community its characteristics. The influence of the pollution will extend even to organisms which are not directly susceptible, as the system accommodates itself to the new conditions. Indeed, it is frequently possible to observe consistent zonation patterns in the biological community which occur in response to sources of pollution. The response to organic pollution in rivers, for example, is particularly well known.

2.3   ORGANIC POLLUTION

The discharge of excessive quantities of organic matter is undoubtedly the oldest, and even today the most widespread form of water pollution. Its significance to human health is discussed in Chapter 5; the present discussion is concerned with its effect on the receiving waters.

The major sources of organic pollution are sewage and domestic wastes; agriculture (especially runoff from inadequately stored animal wastes and silage); the various forms of food processing and manufacture; and numerous industries involving the processing of natural materials such as textile and paper manufacture. Most organic wastewaters contain a high proportion of suspended matter, and in part their effects on the receiving water are similar to those of other forms of suspended solids (see section 2.8). However, the most important consequences of organic pollution can be traced to its effect on the dissolved oxygen concentration in the water and sediments. In an unpolluted water, the relatively small amount of dead organic matter is readily assimilated by the fauna and flora. Some is consumed by detritivorous animals and incorporated into their biomass. The remainder is decomposed by bacteria and fungi, which are themselves consumed by organisms at higher trophic levels. The activity of micro-organisms results in the breakdown of complex organic molecules to simple, inorganic substances, such as phosphate and nitrate ions which become available as nutrients for photosynthetic plants, and carbon dioxide and water. During these metabolic processes, oxygen is consumed. However, where the organic load is light, the oxygen removed from the water is readily replaced by photosynthesis and by re-aeration from the atmosphere.

Where the input of organic material exceeds the capacity of the system to assimilate it, a number of changes take place. How far the sequence of changes proceeds depends upon the severity of the organic load and the physical characteristics of the receiving water. Initially, the enhanced level of organic matter will stimulate increased activity of the aerobic decomposer organisms. When their rate of oxygen consumption exceeds the rate of re-aeration of the water, the dissolved oxygen concentration in the water will fall. This alone may be sufficient, as argued earlier, to eliminate some species, which may or may not be replaced by others with less rigorous demands for oxygen. If the drop in oxygen concentration is very severe, the aerobic decomposers themselves will no longer be able to function, and anaerobic organisms will become predominant.

The biochemical reactions involved in the breakdown of organic matter, and the micro-organisms involved, are described in general terms by Dugan (1972) and by Higgins & Burns (1975). The composition of organic waste varies according to its source, and in particular according to the relative abundance of material of plant, animal or microbial origin. Typically, however, the composition of organic waste is approximately as shown in Table 2.3. Most effluents, of course, also contain other materials, in particular toxic matter (see Tables 2.1 and 2.2), derived from various sources. To illustrate the effects of the breakdown of organic matter on the receiving

Table 2.3 — Approximate analysis of the organic component of sewage (after Higgins & Burns, 1975)

Lipids	30%
Amino acids, starch, proteins	8%
Hemicellulose	3%
Cellulose	4%
Lignin	6%
Protein	25%
Alcohol-soluble fraction	3%
Ash	21%

water, proteins may be used as an example. The first stage of decomposition of proteins is usually their breakdown by hydrolysis to their constituent amino acids. A typical amino acid is alanine, and its breakdown under aerobic conditions may be summarised:

$$CH_3.\overset{\overset{\displaystyle NH_2}{|}}{\underset{\underset{\displaystyle H}{|}}{C}}-COOH + \tfrac{1}{2}O_2 \rightarrow CH_3.\overset{\overset{\displaystyle O}{\|}}{C}-COOH + NH_3$$

        Alanine   Pyruvic acid   Ammonia

Pyruvic acid is an important substance in the metabolism of most living organisms. It is produced in normal metabolism during the glycolytic (anaerobic) phase of the breakdown of carbohydrates and, as in this case, from the breakdown of excess amino acids. In aerobic organisms, the pyruvate enters the citric acid cycle, the primary means by which compounds are broken down to release energy, carbon dioxide and water. Under aerobic conditions, therefore, proteins will be broken down ultimately to these relatively innocuous compounds, while providing a source of metabolic energy for the organisms responsible for the catabolism. Ammonia is also a common end product of the metabolism of nitrogenous compounds (such as amino acids) and in aquatic organisms is generally excreted as such. Normally, the ammonia diffuses rapidly into the environment, but where the level of organic enrichment is high, it can create difficulties for living organisms as it is very toxic (see section 2.7.2). Typically, therefore, organic wastes contain high levels of ammonia; the eventual fate of the ammonia is very relevant to the effects of water pollution and is discussed below.

Under anaerobic conditions, the breakdown of amino acids takes place through different metabolic pathways. Some amino acids, such as cysteine, contain sulphur as well as nitrogen, and its breakdown is used as an example in the following sequence of reactions which are catalysed by acid-producing and methanogenic bacteria:

$$4C_3H_7O_2NS + 8H_2O \rightarrow 4CH_3COOH + 4CO_2 + 4NH_3 + 4H_2S + 8H$$
$$4CH_3COOH + 8H \rightarrow 5CH_4 + 3CO_2 + 2H_2O$$

In this case, then, the products of decomposition include (in addition to ammonia, carbon dioxide and water) acetic acid, hydrogen sulphide and methane. These compounds are very toxic to most forms of aquatic life and, in addition, they are aesthetically undesirable by virtue of their unpleasant odours.

The fate of the ammonia largely depends upon the level of oxygen present. Under aerobic conditions, nitrifying bacteria predominate and the ammonia is converted to nitrite (e.g. by bacteria of the genus *Nitrosomonas*) and subsequently to nitrate (e.g. by *Nitrobacter* spp.). Thus the toxic ammonia is oxidised to the less toxic nitrite and to the relatively innocuous nitrate. Since, however, both ammonia and nitrate are important plant nutrients, problems related to eutrophication can arise as a consequence of organic inputs to water (see section 2.4). Under anaerobic conditions, denitrification of the nitrate typically takes place under the influence of other bacteria such as *Thiobacillus denitrificans* and some *Pseudomonas* species. These cause the reduction of nitrate to elemental nitrogen, which readily displaces the less-soluble oxygen from solution and contributes still further to the deoxygenation of the water. The processes are most easily understood by reference to Fig. 2.1 which summarises the nitrogen cycle in the aquatic environment.

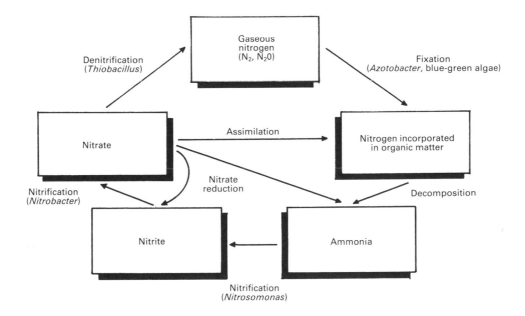

Fig. 2.1 — Summary of the nitrogen cycle in fresh-water ecosystems.

These chemical changes, combined with the blanketing effect of fine organic particles on the substratum, lead to the deoxygenation of the water and substratum, and readily bring about profound changes in the fauna and flora of the receiving water. In severe cases, animal life may be completely eliminated. Hynes (1960) summarised in diagrammatic form the changes which typically occur in a river below a discharge of organic effluent (Fig. 2.2).

The strength of an organic effluent is frequently expressed in terms of its *biological oxygen demand* (BOD). This is defined as the quantity of oxygen utilised, expressed in mg l^{-1}, by the effluent during the microbial degradation of its organic content. BOD is typically measured by taking a sample of water or effluent and aerating it vigorously until it is saturated with oxygen. The dissolved oxygen concentration in part of the aerated sample is then determined by one of the well-known standard procedures. Another part of the aerated sample is incubated, in a sealed bottle of known volume for a period of (typically) five days, at a controlled temperature which is usually 15°C or 20°C. Usually the incubation is carried out in the dark, to eliminate oxygen production by any photosnythetic organisms which may be present. At the end of the incubation period, the concentration of dissolved oxygen remaining in the sample is determined. The difference between the initial and final dissolved oxygen concentrations is used to determine the BOD of the sample, which is expressed as milligrams of oxygen consumed per litre of sample. The basic assumption of the method is that oxygen is mainly consumed by aerobic micro-organisms during the metabolism of organic matter. This is not necessarily true, since many effluents contain chemically-reduced compounds which undergo oxidation by purely chemical reactions. For this reason, it is often necessary to carry out other determinations (for example, of chemical oxygen demand, or of permanganate value) in addition to BOD in order to characterise precisely the likely effects of an effluent or correctly interpret the significance of a BOD value. Within these limitations, BOD values are generally useful as indicators of the organic loading of water. They typically range from one or two milligrams per litre in unpolluted water, to 50 000 milligrams per litre or more in effluents or severely polluted receiving waters. The wide range of expected values means that the precise details of the method of determination have frequently to be modified according to the circumstances; raw samples, for example, often need to be diluted before the determination is carried out. The incubation period of five days (rather than some shorter period) in practice compensates for the fact that some samples will have, initially, very few micro-organisms present, while others will contain a large inoculum. Detailed descriptions of the procedures which may be employed for BOD and other relevant determinations, and for the interpretation of their results, are available in several handbooks, for example APHA (American Public Health Association, 1981).

Fig. 2.2 (A) shows the immediate effect of the effluent on the BOD of the receiving water. Correspondingly, the dissolved oxygen (DO) level drops, gradually recovering as the BOD falls with increasing distance downstream. The oxygen sag curve varies in its dimensions according to the strength of the effluent and the physical characteristics of the river. (Knowing the BOD of the effluent and certain physical characteristics of the river, oxygen sag curves can be predicted according to mathematical models. This can be useful in planning, for example by predicting the likely effect of a proposed new outfall on a receiving water.) Fig. 2.2 (A) also shows the typical changes in the suspended solids and dissolved salt concentrations of a receiving water downstream of an organic effluent. Fig. 2.2 (B) shows how ammonia, nitrates and phosphates typically behave. Ammonia in the receiving water reaches peak levels at the point of greatest deoxygenation, but declines as aerobic conditions are re-established and reduced nitrogen compounds are oxidised to nitrates.

Fig. 2.2 (C) and (D) show the responses of the plant and animal communities to

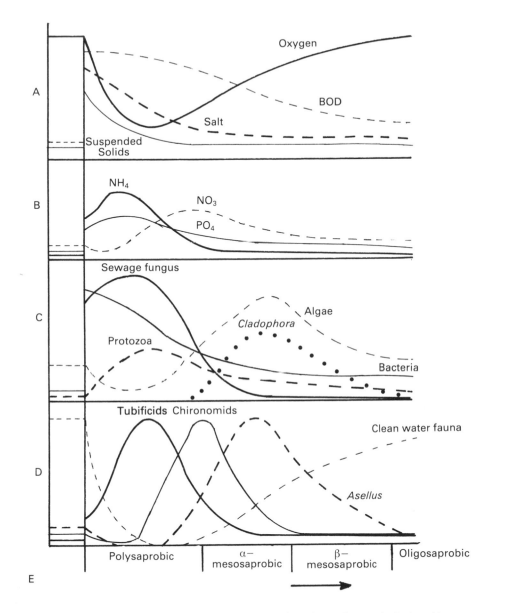

Fig. 2.2 — Changes in water quality and abundance of selected organisms typically found in a river below the discharge of an organic effluent, after Hynes (1960). See text for explanation. Fig. 2.2 (E) shows the zones described by Kolkwitz & Marsson (1909). Other descriptive systems have been used (see Hellawell (1986) for a summary) but are not significantly different.

these physical and chemical changes. 'Sewage fungus' is a characteristic and conspicuous feature of waters which are heavily polluted with organic matter. It is not, in fact, a fungus at all, although its dense, blanketing growths of matted, greyish-brown filaments suggests this idea to many people. It is an aggregation, of varying composition, of bacteria, algae, fungi and protozoa, frequently dominated by the slime-forming bacterium *Sphaerotilus natans* (Curtis & Curds, 1971). Algae are generally reduced initially (partly because the high levels of suspended solids prevent photosynthesis), but increase rapidly in abundance downstream as light penetration improves and levels of nitrate and phosphate remain high. *Cladophora* is a particularly conspicuous attached filamentous alga, widely associated with mild organic pollution or with the early stages of recovery from more severe pollution, and appears to be especially responsive to elevated phosphate levels (Whitton, 1970; Pitcairn & Hawkes, 1973).

The animal community (Fig. 2.2D) also shows a clear pattern of response. The 'clean-water fauna' initially declines, and may be entirely eliminated, in the region immediately below the outfall. Tubificid worms, being typically tolerant of low dissolved oxygen levels and silty substrata, usually dominate the fauna in the more seriously affected areas. The larvae of the midge *Chironomus* (Diptera) spp. typically become established further downstream, followed by the isopod crustacean *Asellus,* and the gradual re-establishment of the 'clean-water fauna' as the river returns to its normal physical and chemical status. In any particular situation, the total number of species involved is very large. The general pattern, subject to many variations in detail, is of a zonation, with increasing distance from the source of pollution, every bit as obvious as that along the length of an unpolluted river, or along a transect on a steeply-sloping sea shore. The distribution of many individual species in response to organic pollution has been studied in detail in many different parts of the world. Hellawell (1986) gives a detailed summary of these observations. Several early investigators attempted to provide systematic descriptions of the zonation pattern of plants and animals observed in response to organic pollution with increasing distance from the pollution source, and one such description is shown in Fig. 2.2E. These descriptive systems eventually gave rise to the idea of indicator organisms, and to the development of numerical indices (pollution indices and biotic indices), which are widely used in the biological surveillance of water quality. These topics are further discussed in Chapter 3.

2.4 NUTRIENT POLLUTION

Plant growth in water may be limited by any of several factors, including light and the physical characteristics of the habitat. In many cases, however, the limiting factor is the availability of inorganic nutrients, particularly phosphate (Moss, 1980). Increased input of nutrients can therefore trigger increased plant growth which, if excessive, leads to changes in the biological characteristics of the receiving water. The discharge of organic matter to water is an important source of plant nutrients, since the aerobic decomposition of organic matter results in the release of phosphate, nitrate and other nutrients. Domestic sewage typically contains high levels of phosphate because detergent washing powder formulations normally contain high levels of phosphate. For example, the level of phosphate typically found in treated sewage effluent (Table 2.1) may be compared with the levels normally found in unpolluted waters, which range from about 0.001 to 1 mg l^{-1} (Moss, 1980). Food-processing effluents are often high in nitrate and phosphate, and in agricultural areas runoff from land carries nutrients into the water, especially if artificial fertilisers are used. Many agricultural and forestry practices lead to increased soil erosion, carrying plant nutrients from the land to the water. Intensive rearing of livestock contributes significant nutrient loads to surface waters.

Increased plant growth can sometimes be considered beneficial, especially in oligotrophic waters where primary productivity is nutrient-limited. Moderately-increased plant growth can provide increased productivity of herbivorous and detritivorous animals, leading to increased overall productivity. It is not unknown, for example, for fishermen deliberately to 'fertilise' lakes to increase fish yield. The increased spatial heterogeneity of the habitat can also give rise to an increase in species diversity. Excessive plant growth, however, causes four main adverse consequences. The blanketing effect of macrophytes and filamentous algae can result in major faunal alterations owing to physical changes in the habitat. Respiration of dense plant growths can produce depressed dissolved oxygen levels, not only at night when photosynthesis ceases but also during the day if the density of plant growth reduces light penetration. Some algal species, under the influence of elevated nutrient levels, 'bloom' — that is they reproduce rapidly and dominate the flora. These algal blooms give rise to several problems, including tainting and discoloration of the water (rendering it unsuitable for potable supply) and the production of toxins which are harmful to fish and invertebrates. Finally, the eventual decay of the plant biomass has exactly the same effect as the input of a large quantity of allochthonous organic matter.

2.5 EUTROPHICATION

The phenomenon of eutrophication is particularly associated with lakes and slow-flowing waters. It is widely, and erroneously, believed that pollution by plant nutrients and organic matter actually causes eutrophication. It is more accurate to say that pollution accelerates what is probably a natural process. To understand the causes and consequences of eutrophication requires some knowledge of the special characteristics of lakes.

In temperate latitudes, most lakes were formed by glaciation. Moving glaciers gouged out hollows in the earth, and when the ice retreated these hollows became filled with water from the melting ice. Lakes are not, therefore, geologically ancient phenomena. In modern times, substantial man-made lakes have become common in many parts of the world. A lake is a body of water which is very slow-moving. Some lakes have rivers flowing into or out of them. Even those which do not, however, are not static; water moves slowly into or out of the lake via the ground. Because the water moves only very slowly, some physical and chemical processes occur in lakes which do not occur in moving waters. Of particular importance are stratification, and temporal variations in chemical quality of the water.

Stratification occurs because the lake water is heated by the sun at the surface. Because warm water is less dense than cooler water, and water is a poor conductor of heat, during the warmer months of the year an upper layer of warm water, the epilimnion, becomes established and sharply delineated from a lower layer of cooler water, the hypolimnion. Between them is a very narrow zone, the thermocline, within which the water temperature drops very sharply with only a slight increase in water depth (Fig. 2.3). Little or no vertical mixing can take place, the lake being

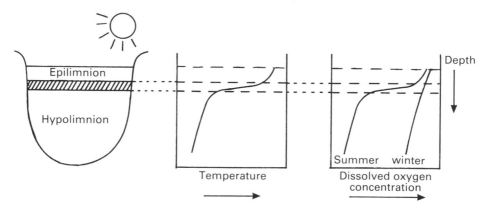

Fig. 2.3 — Thermocline formation in a typical lake. The diagram on the left shows the stratification which develops, the thermocline being represented by the shaded area. The centre diagram shows the temperature profile. The diagram on the right shows the deoxygenation of the hypolimnion which occurs in eutrophic lakes in the summer. The vertical mixing which occurs in winter tends to abolish the stratification, redistributing oxygen and plant nutrients through the water column.

effectively divided horizontally into two distinct layers separated by the thermocline. Obviously, stratification cannot occur in very shallow lakes.

Photosynthesis can only occur in shallow water, where light can penetrate. At the lake margins, emergent plants and rooted aquatic macrophytes occur, but as the depth of the water increases, primary production is possible only by phytoplankton in the surface waters, within the epilimnion. During the winter, phytoplankton growth is restricted by low temperatures and low light intensity. In spring and summer increasing temperatures and light intensity stimulate phytoplankton growth, leading to an increase in population density and the depletion of dissolved nutrients in the water of the epilimnion. Plant growth and reproduction slow down, and as the plant cells senesce and die, they sink into the hypolimnion and eventually to the bottom of the lake, where they begin to decompose. The inorganic nutrients which are the products of decomposition remain in the hypolimnion, however, as the stratification prevents vertical mixing of the water and upwards diffusion is slow. As autumn approaches, reduced temperatures, light intensity and limited nutrients accelerate the decline of the phytoplankton population. In the winter, the epilimnion cools and becomes more dense. Its water sinks, displacing the hypolimnion which is now warmer and lighter than the epilimnion. The lake waters become thoroughly mixed, and nutrients from the hypolimnion are brought to the surface, bringing about conditions suitable for the start of the next annual cycle (Fig. 2.4).

Underlying these annual cycles is a progressive change in the physical and chemical characteristics of the lake. At its formation, the lake contains few plant nutrients or dissolved minerals of any kind, and a negligible quantity of organic matter. With the passage of time, dissolved minerals including plant nutrients enter the lake from surface runoff and groundwater infiltration, at a rate which depends largely upon the climate and the geology of the surrounding area. As the nutrient levels rise, a flora and fauna becomes established and develops, contributing an

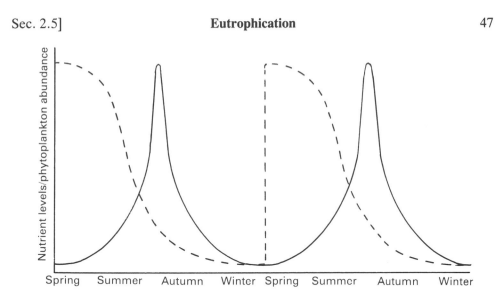

Fig. 2.4 — Seasonal cycling of nutrients and phytoplankton in the surface waters of a lake. Nutrient levels (dotted line) decline as phytoplankton growth (solid line) takes place. Stratification leads to nutrient depletion of the surface layers, so the phytoplankton population declines and senescent cells fall to the hypolimnion. Abolition of the thermocline in winter redistributes the plant nutrients allowing the cycle to begin again the following spring.

increased content of organic matter in the lake. Organic matter is also gradually accumulated from outside the lake, progressively building up a layer of sediment on the lake bottom. Airborne dust also falls into the lake, and the lake begins slowly to fill up. The rate at which this happens varies from the barely detectable up to a few millimetres per year. The gradual deposition of material on the floor of the lake basin causes the lake to shrink, new land being formed at its edges (Fig. 2.5). This new land is colonised by terrestrial plants, and in some lakes it is possible, by walking away from the lake's edge, to see clearly the various stages of development of the terrestrial flora, a classic example of ecological succession. In areas where these processes have occurred, for various reasons, at different rates in different lakes, it is possible to see contemporaneously all the stages of a lake's development from nutrient-poor, sparsely-populated lakes of low productivity, through various stages of nutrient enrichment, to swamp or marsh and eventually dry land.

The term *eutrophication* is applied to the process whereby the nutrient levels of lakes increase from oligotrophic (nutrient poor) to eutrophic (nutrient rich). It appears to be a natural process, although some authors have argued that it is not inevitable or intrinsically unidirectional (Moss, 1980). Since its basic cause is, however, the accumulation of plant nutrients and organic matter in the lake basin, clearly anthropogenic influences will accelerate it. The transition from oligotrophic to eutrophic is accompanied by qualitative and quantitative changes in the biota. Since plant growth is commonly limited by nutrient levels, a gradual increase in nutrient levels would be expected to lead to successional changes in the plant community and corresponding changes in the animal community. Animals, in particular, are likely also to be affected by deoxygenation of the hypolimnion. In

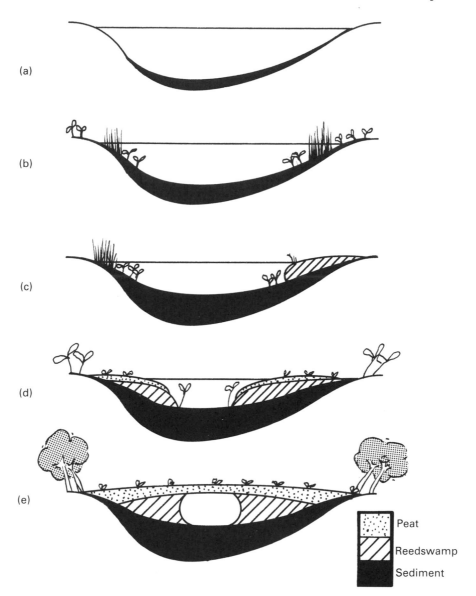

Fig. 2.5 — Stages of development of a typical glacially-formed temperate lake from oligotrophic (upper diagrams) to eutrophic (lower diagrams).

eutrophic lakes, the stratification which leads to nutrient depletion of the epilimnion
also causes oxygen depletion of the hypolimnion. Oxygen demand due to aerobic
decomposition of detritus is high in the hypolimnion, but the absence of either
vertical mixing or photosynthesis in the hypolimnion prevents re-oxygenation of the
hypolimnion from the atmosphere (Fig. 2.3). The sensitivity of aquatic animals to
dissolved oxygen levels, and its consequences for their patterns of distribution and
abundance, has already been discussed.

Moss (1980) and Mason (1981) discuss in detail some well-studied case histories of eutrophication. Attempts have been made to restore eutrophic lakes, with varying degrees of success, and these are briefly discussed by Mason (1981).

2.6 THERMAL POLLUTION

The extent of the use of water for cooling is formidable. Howells (1983) estimated that in Britain 28% of rainfall and 50% of river flow were utilised for this purpose. In fact, Britain has relatively few large inland power stations, and the bulk of the demand falls upon a single river system, the Trent (Lester, 1975) which supplies cooling water for approximately one-third of the nation's power generating capacity; most of the remainder is situated on coastal or estuarine sites. In Britain, direct cooling is rare; water abstracted from the river is recirculated within the plant and much of the excess heat is dissipated through cooling towers. Direct cooling would in fact require far more water than is available from the relatively small rivers. In the USA, direct cooling is more common, and about 10% of runoff is used for cooling (Castenholtz & Wickstrom 1975). Very large rivers are, of course, more common in large continental landmasses; further, the climate in much of the USA is such that river temperatures of 30°C or more are not uncommon, whereas in Britain river temperatures above 24°C are extremely rare. To this extent, the problems of dissipating waste heat differ widely from one location to another.

Temperature is of such profound importance in chemical and biological processes that the effects of temperature alterations on aquatic biological communities is potentially large. Hot effluents from industrial processes and power generation can cause temperature increases in the receiving water of 10°C or more. Some effluents, such as water pumped from deep mines or regulating reservoirs, may be significantly colder than the receiving water, although the effects of cold effluents have received relatively little attention. Because the density of water alters with temperature, hot effluents often form a surface plume rather than mixing quickly with the receiving water. This can exacerbate some of the adverse effects, but may sometimes act to minimise the influence of the effluent on the benthic community, and fish can avoid the elevated temperature by remaining in deeper water.

Elevated temperatures can influence aquatic organisms directly, as the organisms respond physiologically or behaviourally to the new conditions; or indirectly, as the changed temperature influences the chemical environment. For example, increased temperature reduces the solubility of oxygen in water. At the same time, it may increase BOD by stimulating more rapid breakdown of organic matter by micro-organisms. Temperature affects the toxicity of some poisons (see Chapter 4) either through its effect on the organisms themselves or because the dissociation of ionisable pollutants (such as ammonia or cyanides, see section 2.7.2) is temperature-dependent. The direct effects of elevated temperature on fishes have been particularly well-studied and may be used to indicate the potential impact of thermal pollution on aquatic animals generally. Varley (1967) gives a very readable introduction to the relationships between fish and their thermal environment, and their consequences for fish distribution patterns. Alabaster & Lloyd (1980) provide a detailed review of the literature relating to the temperature requirements of freshwater fishes, and Hellawell (1986) provides a concise summary of the effects of thermal pollution on the aquatic environment.

The maximum temperature which fish can withstand varies from species to species, and also within a species according to the environmental history of the fish. Generally, fish can acclimate to gradually-rising temperatures, so that the lethal temperature depends to some extent on the temperature to which the fish was initially acclimated. Relatively small, sudden changes of temperature which do not allow the acclimation process to occur can be more harmful than larger, more gradual changes. Acclimation to altered temperature is probably achieved by the induction and synthesis of isoenzymes. Many enzymes are known to exist in several forms, each having the same function but each modified to perform optimally at a particular temperature, or under some other specific condition. This is one way in which poikilothermic animals can continue to function over wide ranges of temperature, although the adjustments involved require periods ranging from several hours to several days. The relationship between acclimation temperature and the upper lethal temperature may be summarised in a temperature tolerance diagram (Fig. 2.6). The temperature at which the upper lethal temperature ceases to rise with

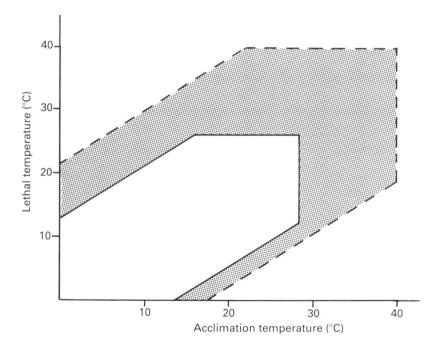

Fig. 2.6 — Temperature tolerance diagrams for (dotted line) a eurythermal warm-water-adapted fish and (solid line) a stenothermal cold-water-adapted fish. The upper lethal temperature increases with acclimation temperature until the ultimate upper lethal temperature is reached. The lower lethal temperature is also influenced by acclimation temperature. Only temperature changes within the boundaries are tolerated.

acclimation temperature is termed the *ultimate upper lethal temperature*. Values range from about 24°C for Salmonid fishes to 40°C or more for fishes characteristic of warmer environments. In general, the ultimate upper lethal temperature for a particular species is several degrees higher than any temperature likely to be encountered in its normal habitat. Therefore death of fish due to high temperature alone is probably rare, and the adverse effects of elevated temperature are due to more subtle mechanisms.

The effects of temperature on the respiratory physiology of fish are particularly important. Even under favourable conditions, aquatic animals face formidable difficulties in balancing their needs for oxygen against the meagre quantity available from water and the high energy cost of obtaining it (see section 2.2). Increased temperature both reduces the amount of oxygen available and increases the animal's demand for it. In addition, for animals in the wild it is not sufficient to survive passively; survival presupposes the need to engage in physical activity, which imposes still further demands for oxygen. In fish, constraints imposed by the size, structure and efficiency of the gills, and by the limited solubility of oxygen in water, limit the maximum possible oxygen consumption to a value of about $400 \, \text{mg kg}^{-1} \text{h}^{-1}$ in most species. In Fig. 2.7, the effect of temperature on the oxygen consumption of

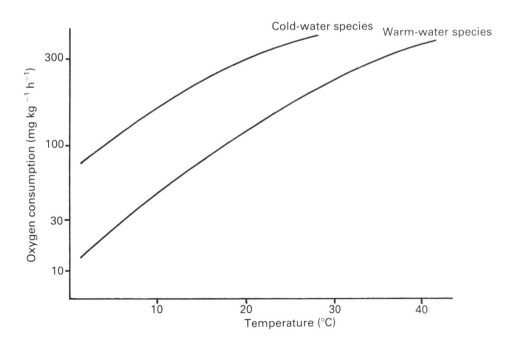

Fig. 2.7 — Relation between oxygen consumption and temperature for fish at rest. The upper line represents a cold-water-adapted species, the lower line a warm-water-adapted species.

two fish species is shown. Warm-water species require to reach the maximum possible rate of oxygen expenditure only at fairly high temperatures. As the temperature falls, their metabolic rate falls until it is too low to permit physical activity, and the fish become torpid. In the wild, they would not survive because they would be unable to find food, avoid predators, or even maintain their position in a current. Cold-water species, such as Salmonids, however, must be active at low temperatures. They are adapted to high rates of oxygen consumption at low temperatures. However, their rate of oxygen consumption increases with temperature at about the same rate as that for cold-water species (Fig. 2.7). Therefore they reach the maximum possible rate at a much lower temperature. As argued in section 2.2, when the fish reaches this condition the energy cost of working the respiratory muscles is such that there is no oxygen available for the tissues, and the fish dies. Further, the need to retain a margin of capacity for physical activity is important. At its ultimate upper lethal temperature, the fish has no scope to indulge in physical activity other than respiratory movements. At very low temperatures, it has little scope for muscular activity of any sort, since the efficiency of muscle tissue decreases with temperature. At intermediate temperatures, the difference between basal metabolism (oxygen consumption at rest) and active metabolism (oxygen consumption during physical activity) reaches a distinct maximum (Fig. 2.8). In practice, the

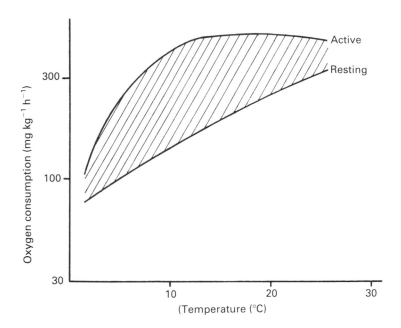

Fig. 2.8 — Effect of temperature on (lower line) the resting or standard metabolism of a fish and (upper line) the active metabolism. The shaded area represents the 'scope for activity' of the fish.

temperature regime which is favourable for the indefinite survival of the fish includes a much narrower range of temperatures than that which would allow the survival of the fish under laboratory conditions.

The range of temperatures which is suitable for growth, reproduction and development of fishes is also generally rather narrower than that which simply allows survival. Temperature is, in conjunction with photoperiod, an important trigger for the onset of reproductive cycles. Temperature also governs the rate of development of fish eggs. In most species, the average temperature multiplied by the time taken for the eggs to hatch is a constant. In Salmonids, eggs typically require 410 degree-days; for example, 41 days at 10°C or 51 days at 8°C. This relationship only holds good, however, within a certain temperature range — below about 5°C and above about 15°C, the proportion of eggs successfully developing is markedly reduced. In Cyprinid fishes, development of eggs may be much quicker, typically three days at 20°C in carp, with other species showing intermediate values. Thus quite short-term anomalies in the temperature regime, if they occur at a critical period, can exert a serious effect on fish reproduction. Studies of the growth of many fish species, particularly those which are widely cultured, show that temperature has critical effects on growth rate and on food conversion efficiency (i.e. the ratio between the increase in weight of the fish and the quantity of food consumed). (In fish, sexual maturity normally occurs when the animal has reached a critical size rather than a critical age.) Clearly, temperature anomalies can have a major influence on the reproductive success of fish, without necessarily giving rise to any obvious immediate effects on the adult population. In practice, fish have considerable powers of mobility and in laboratory experiments show clear preferences for particular temperatures, ranging from about 13°C for Salmonids to 37°C for carp (Varley, 1967). Thus areas subject to thermal anomalies may be devoid of fish through avoidance reactions, but without any serious long-term effects on the population, provided that the fish do have access to cooler water.

In view of the prevalence of heated effluents, and of the potential importance of elevated temperatures indicated by laboratory studies, it is perhaps surprising that unequivocal examples of ecological damage by thermal pollution are rare. Recent reviewers of the topic, for example Castenholtz & Wickstrom (1975), Mason (1981), Howells (1983), and Hellawell (1986), have cited no clear instances of any readily-detectable adverse effect of elevated temperature, as such, on the ecology of a receiving water, apart from numerous autecological studies whose significance to the overall ecology of the system is unknown. However, the review of Alabaster & Lloyd (1980), based largely upon research (including field studies) in eastern Europe which was previously little-known, suggests that temperature increases of between 2 and 8°C do produce significant alterations in the biota of receiving waters. The apparent difficulty of detecting the ecological effects of thermal pollution may be due to any of several causes. In eastern Europe, where such effects are more apparent, the prevailing climate is significantly cooler than in western Europe or the USA where most studies have been carried out. In many circumstances, heated effluents are discharged to rivers which are also polluted with organic or toxic matter, and the effects of elevated temperature may be difficult to distinguish from other pollution effects. Local conditions undoubtedly influence the impact of heated effluents. In many countries, power generation reaches its maximum during the winter months when river temperatures are low and discharges are high. In others, depending upon climatic and economic factors, demands for electric power (for refrigeration or air-conditioning) in summer, when river discharges are lowest, may approach or exceed winter levels. The effects of heated effluents on the ecology of receiving waters may therefore be expected to vary widely from one region to another. Finally, the effects of elevated temperatures may be difficult to predict or detect without detailed knowledge of specific local circumstances. For example, increased temperature will accelerate the microbial metabolism of organic matter. In sluggish, poorly-aerated rivers this will accentuate the effects of organic pollution; but in turbulent waters which re-aerate rapidly from the atmosphere, the effect of elevated temperature would be to reduce the extent of the zone within which the adverse effects of organic inputs manifest themselves, and in a lightly-polluted water could lead to a beneficial increase in overall productivity. It is therefore unwise to attempt generalisations on the effects of thermal pollution; each case must be considered individually.

2.7 TOXIC POLLUTION

There are about four million different chemical substances in existence, a number which increases by about 300 000 every year. Of these, about 63 000 are in common use (Maugh, 1978). Goodman (1974) estimated that about 10 000 chemicals are produced in quantities exceeding 500 kg yr^{-1}. A large proportion of these thousands of chemicals are, presumably, only produced and/or used in only a small number of locations. Nevertheless the number of pollutants which can be considered as widespread is still formidably large. A realistic figure is indicated by the 1978 Great Lakes Water Quality Agreement between the US and Canadian governments (reprinted in Nriagu & Simmons, 1983). Appendix 1 of this agreement lists 271 different substances which, on the basis of toxicological and discharge data, are considered hazardous to the North American Great Lakes. Appendix 2 of the same agreement lists a further 106 'potentially hazardous polluting substances'.

The present discussion is confined, of necessity, to a general description of the sources and characteristics of some of the major categories of toxic pollutant. The categories chosen are for convenience, and are not necessarily mutually exclusive; many pesticides, for example, contain heavy metals. It is not at all easy to give specific descriptions of the effects of a given pollutant, because of the diversity of the chemicals involved, the wide range of environmental conditions which prevail in different receiving waters, the fact that poisons frequently occur in complex mixtures, and the enormous differences in the physiology of the organisms which are exposed to them. The study of pollutant toxicity and toxic effects is described in some detail in Chapters 1 and 4. Specific information on the toxicity and toxic effects of pollutants is best sought in review articles and other data compilations, which are referred to where appropriate. A very comprehensive recent reference work is that of Hellawell (1986). It is important to note, however, that toxicity data obtained from such compilations can be extremely misleading, for reasons which are explained fully in Chapter 4. It is recommended that wherever possible the original sources are consulted, and the validity of the data assessed in relation to the methodological criteria described in Chapter 4.

2.7.1 Heavy metals

'Heavy metals' is an imprecise term that is generally taken to include the metallic elements with an atomic weight greater than 40, but excluding the alkaline earth metals, alkali metals, lanthanides and actinides. The most important heavy metals from the point of view of water pollution are zinc, copper, lead, cadmium, mercury, nickel, and chromium. Aluminium (see section 2.9) may be important in acid waters. Some of these metals (e.g. copper, zinc) are essential trace elements to living organisms, but become toxic at higher concentrations. Others, such as lead and cadmium have no known biological function. Industrial processes, particularly those concerned with the mining and processing of metal ores, the finishing and plating of metals and the manufacture of metal objects, are the main source of metal pollution. In addition, metallic compounds are widely used in other industries: as pigments in paint and dye manufacture, in the manufacture of leather, rubber, textiles and paper, and many others. Quite apart from industrial sources, domestic wastes contain substantial quantities of metals because the water has been in prolonged contact with copper, zinc, or lead pipework or tanks. Some forms of intensive agriculture give rise to severe metal pollution; copper, for example, is widely added to pig feed and is excreted in large quantities by the animals. Mance (1987) gives a detailed review of metal pollution.

Heavy metals may be classed (see Table 2.4) generally as toxic or very toxic to

Table 2.4 —Degrees of pollutant toxicity, classified according to the scheme proposed by the Joint IMCO/FAO/UNESCO/WHO Group of Experts on the Scientific Aspects of Marine Pollution (1969)

Degree of toxicity	Acute toxicity threshold, $(mg\,l^{-1})$
Practically non-toxic	Above 10 000
Slightly toxic	1000–10 000
Moderately toxic	100–1000
Toxic	1—100
Very toxic	Below 1

aquatic animals and to many plant species, though large interspecific differences in susceptibility occur even within closely-related groups of organisms. Relatively little is known about the interaction of heavy metals with the aquatic microbial flora. Algae, macroinvertebrates and fishes have, however, been widely studied. In general, the heavy metals may be listed in approximate order of decreasing toxicity as follows: Hg, Cd, Cu, Zn, Ni, Pb, Cr, Al, Co. However, this sequence is very tentative and the position of each element in the series will vary with the species tested and the conditions of the experiment. Apart from some remarkable interspecific variations in susceptibility to metals, the toxicity of most metals varies enormously with the environmental conditions, mainly because of the effect of environmental conditions on the chemical speciation of the metal (see section 4.2). Study of the ecological effects of heavy metals as water pollutants is often hampered by the fact that other pollutants are normally additionally present. However, there is an enormous literature on heavy metal toxicity (see, for example, Whitton & Say, 1975; Alabaster & Lloyd, 1980; Hellawell, 1986) and some important aspects of the impact of heavy metals as water pollutants have already been discussed in Chapter 1.

Two features of heavy metal toxicity which should not be overlooked are their ability to form organometal complexes and their potential for bioaccumulation. There is some evidence (see section 4.2) that the presence of organic substances can reduce heavy metal toxicity considerably, at least as measured in conventional toxicity tests. However, a number of organometal compounds are known to be particularly hazardous to aquatic life. Tributyl tin, for example, a constituent of anti-fouling paints, is implicated in severe environmental damage in harbours, boatyards and inland waters, and appears in the 'Black Lists' of substances compiled by international organisations such as the EEC and United Nations Environment Programme. Similarly, the dangers associated with methylated forms of mercury are well known. Many metals, whether organically-complexed or not, are known to accumulate in plant and animal tissues to very high levels, posing a potential toxic hazard to the organisms themselves, or organisms higher in the food chain including humans, which may consume them.

2.7.2 Ammonia, cyanides and phenols

Ammonia, cyanides and phenols are considered together because, with copper and zinc, they are the most widespread and serious toxic water pollutants in industrialised countries. Ammonia and its compounds are ubiquitous constituents of industrial effluents because ammonia is a staple raw material in many branches of the chemical industry; it is, therefore a common end product of industrial processes as well as an important by-product of others, notably the production of coke and gas from coal, from power-generation and from most processes involving the heating or combustion of fuel. It is also a natural product of the metabolism of organic wastes in treatment plants and receiving waters. The toxicity of ammonia to fish is well documented, and although less is known of its effect on invertebrates it appears that levels of ammonia which are tolerable to fish present little danger to most invertebrates (Alabaster & Lloyd, 1980). In aqueous solution, ammonia forms an equilibrium between unionised ammonia, ammonium ion and hydroxide ions:

$$NH_3 + H_2O \rightleftharpoons NH_4^+ + OH^-$$

Unionised ammonia is very toxic to most organisms, but ammonium ion is only moderately toxic. The toxicity of the solution therefore depends on the quantity of unionised ammonia. This in turn depends upon the pH and temperature of the water; as pH and temperature rise, the proportion of unionised ammonia also rises. The effect of pH and ammonia on toxicity is therefore considerable. In order to know whether a given level of total ammonia is likely to be toxic, it is necessary to use the pH and temperature values to calculate the corresponding level of free ammonia. (Some authors, such as Alabaster & Lloyd (1980) and Hellawell (1986), have published tables or nomographs to facilitate the calculation.) As an example, the European Inland Fisheries Advisory Commission (EIFAC) recommend that unionised ammonia concentrations should not exceed 0.025 mg l^{-1}. In a water of pH 8.5 at a temperature of 20°C, this corresponds to a *total* ammonia concentration of 0.22 mg l^{-1}. In a cooler, acid water, however (pH 6.5, 5°C), a concentration of *total* ammonia of 63.3 mg l^{-1} would be acceptable, whereas at pH 6.5 and 20°C, the maximum acceptable concentration of *total* ammonia would be 20 mg l^{-1}.

Cyanide is also a very common constituent of industrial effluents, being produced from processes involving coking and/or combustion such as steelworks, gas production and power generation. Cyanides are also used in the hardening, plating and cleaning of metals. Cyanides dissociate in aqueous solution:

$$KCN + H_2O \rightleftharpoons H^+ + K^+ + HCN \rightleftharpoons CN^- + H^+.$$

The dissociation, and consequently the toxicity of cyanide, is pH-dependent, low pH favouring the formation of undissociated HCN which is highly toxic. Cyanide ions readily form complexes with heavy metal ions. The stability and toxicity of these complexes varies according to the metal and also with the pH. Thus the toxicity of cyanides and effluents containing cyanides (which commonly contain substantial amounts of heavy metal) is, as with ammonia, greatly influenced by pH, but is less well understood. It is clear, however (see Chapter 1), that the contribution of cyanides to the overall toxicity of complex industrial effluents can be identified, and is often significant.

Phenolic substances include the monohydric phenols (phenol, cresols and xylenols) and the dihydric phenols including catechols and resorcinols. They are found in a wide range of industrial effluents, and are particularly associated with gas and coke production, the refining of petroleum, power generation, many branches of the chemical industry and the production of glass, rubber, textiles and plastics. Alabaster & Lloyd (1980) and Buikema *et al.* (1979) provide detailed reviews of the literature. The latter review includes discussion of phenol derivatives which are used for specialised purposes such as pesticides. Apart from these, phenolic substances rarely occur as pollutants except as components of complex effluents which contain a variety of other polluting substances. In practice, therefore, the major concern is to determine the extent to which phenols contribute to the overall toxicity of an effluent, and an example was discussed in Chapter 1.

2.7.3 Pesticides
Pesticides are a diverse group of poisons of widely-varying chemical affinities, ranging from simple inorganic substances to complex organic molecules. Of the latter, some are natural metabolites, particularly of plants, while others are synthetic derivatives of natural products or completely synthetic substances produced in chemical factories under conditions which do not exist in the natural world. They have in common only that each pesticide is highly toxic to some forms of life and of intermediate or negligible toxicity to others, and that they have been widely introduced into the natural environment. Pesticides are introduced into aquatic systems by various means: incidentally in the course of their manufacture, and through discharge consequent upon their use. Surface water runoff from agricultural land and the side-effects of aerial spraying are especially important, and many serious pollution incidents arise through the accidental or negligent discharge of

concentrated pesticide solutions which have been used for agricultural purposes such as sheep-dipping. Additionally, many pesticides are deliberately introduced into water bodies to kill undesirable organisms such as insect or molluscan vectors of human diseases; weeds; fish, and algae.

Pesticides are also used in many industrial processes, for example in the manufacture of textiles and in the production and processing of perishable materials such as paper and timber products. They are therefore an important component of many industrial effluents. Because of their enormous diversity and their importance as pollutants, pesticides have attracted an enormous literature. Useful introductions are given, for example, by Khan (1977), Perring & Mellanby (1977) and Brown (1978). A valuable summary of pesticides as water pollutants is given by Hellawell (1986). In many countries, the special significance of pesticides as pollutants (and as widely-used toxic chemicals in the working environment) has led to the development of strict controls on their use. One example is the Pesticides Safety Precaution Scheme operated in Britain by the Ministry of Agriculture.

Insofar as it is possible to generalise about the polluting effects of such a diverse group of substances, the following points are perhaps of greatest significance. Firstly, effective pesticides are more or less selective in their effects, that is they are extremely toxic to some forms of life and relatively harmless to others. Secondly, their modes of application vary according to the circumstances. In some cases, pesticides are applied in relatively high concentrations for relatively short periods. This pattern of application typically occurs where pesticides are applied to water to kill weeds, disease vectors or other undesirable organisms, or as an incidental effect of aerial spraying of crops. Here, the principal concern may be to determine their short-term toxicity to non-target organisms, and it may be advantageous to devise specific toxicity testing protocols in order to estimate the impact of the pesticides on the receiving-water biota (see section 4.1.5). However, in lowland rivers draining agricultural areas, pesticides are more likely to be present at low but fairly consistent levels, and in this case the major areas of toxicological interest will be their potential sublethal effects, their capacity to accumulate in individual organisms and via the food chain, and the development of resistance through acclimation and/or genetic adaptation. Many pesticides are known to be refractory to chemical and biological degradation, and their persistence in the environment has for many years been a cause for concern. Probably the best example of this is the well-known case of DDT, which has been used in such enormous quantities in the last 40 years that no part of the world is now free from measurable contamination, and its manufacture and use in many countries is now banned or severely restricted. One consequence of this, however, is that other pesticides are used as substitutes; many of these are of much more recent discovery than DDT, and may eventually be found to be equally or more dangerous when sufficient information about them has been accumulated.

2.8 SUSPENDED SOLIDS

Virtually all effluents contain suspended particulate matter, but especially those associated with mining and quarrying for coal, china clay, stone and other mineral materials. Dredging, engineering works and boat traffic commonly introduce particulate matter into suspension. Stormwater drainage and surface water runoff also contribute substantial loads. Suspended matter may be organic or inert, and some forms are chemically reactive (for example, the ferric hydroxide precipitate associated with acid mine drainage (see section 2.9). The present discussion is confined to the physical effects of suspended solid matter. The effects of suspended matter on the receiving water biota are both direct and indirect. Direct effects include physical abrasion of body surfaces, and especially of delicate structures such as gills. Physical damage of this kind interferes with respiration and renders the animals susceptible to infections. High levels of suspended particulates may interfere with the filter-feeding mechanisms of invertebrates, and possibly with the feeding of fish which locate their food visually. In laboratory experiments high levels of suspended solids can kill fish. The concentrations of suspended solids at which these effects occur vary with the species and the nature of the particulate matter. Herbert & Merkens (1961) found that kaolin and diatomaceous earth in suspension at a concentration of 270 mg l^{-1} caused substantial mortality to trout over a period of 10–15 days, but other investigations have shown large differences between the lethal concentrations of different types of suspended matter (Alabaster and Lloyd, 1980).

Indirect effects are mainly due to increased turbidity and the blanketing effect of the particulates when they eventually settle. Increased turbidity will reduce or prevent photosynthesis, leading to a reduction in primary productivity or the complete elimination of plants. (Alternatively, certain forms of silting can, depending on the physical conditions, bring about major changes in the community by promoting the formation of stable weed beds.) Salmonid fishes require aerated gravel beds for egg-laying sites, and the silting of gravel beds can eliminate salmonid populations by depriving the fish of suitable nest sites. In many rivers, salmonid population density is governed by the availability of nest sites rather than by biological or environmental factors. Invertebrate distribution patterns are profoundly influenced by the size of the particles composing the substratum (Hynes, 1970). Mayfly nymphs with exposed abdominal gills (e.g. *Ecdyonurus, Rhithrogena, Ephemerella*) may be replaced by species with covered gills or which are adapted for burrowing (e.g. *Caenis, Ephemera*). Insects which crawl upon the substratum may generally be disadvantaged in favour of species whose means of locomotion is better suited to a soft substratum, such as leeches, oligochaetes and some molluscs. Thus the input of even inert fine particulates can readily bring about major community changes. Herbert *et al.* (1961) described the effects of China clay wastes on the ecology of trout streams, and discussed the mechanisms by which alterations occurred. Other field and laboratory studies are reviewed by Alabaster (1972) and Alabaster & Lloyd (1980). The majority of unpolluted British rivers contain fewer than 50 mg suspended solids per litre of water, and about half have under 30 mg l^{-1}. Alabaster & Lloyd (1980) tentatively suggest that waters containing more than 80 mg l^{-1} are unlikely to support good fisheries.

2.9   EXTREME pH AND ACIDIFICATION

Many effluents, especially if untreated, are strongly acidic or alkaline. All natural waters have some buffering capacity, that is the ability to absorb acid or alkaline inputs without undergoing a change in pH. This buffering capacity is usually expressed in terms of the acidity (ability to neutralise alkalis) and alkalinity (ability to neutralise acids) of the water, and is determined by titration in the presence of a suitable indicator. The relationship between pH, acidity and alkalinity of a water is not simple. Acid waters (pH<7) can have measurable alkalinity, and alkaline waters (pH>7) can have measurable acidity. Where the buffering capacity of the water is exceeded by the input of an effluent, the pH of the water will change. Unpolluted natural waters show a pH range from 3.0 to 11 or more; those lying between 5.0 and 9.0 generally support a diverse assemblage of species and this range may be considered broadly acceptable (Alabaster & Lloyd, 1980). This does not mean, however, that pH changes within the range 5.0–9.0 are of no consequence. For example, pH is an important determinant of the distribution patterns of aquatic species, as can be seen from the study of invertebrate communities in unpolluted rivers (Sutcliffe & Carrick, 1973; Sutcliffe, 1983; Haines, 1981). In addition, apparently small changes in pH can have major effects on the toxicity of pollutants such as ammonia (see section 4.2) so that the effect of a given level of pollutant can vary, depending upon pH, from being scarcely noticeable to being extremely serious.

A very common form of pollution involving extreme pH is *acid mine drainage*. Coal mines are the most common source of acid mine drainage, but it can occur wherever mineral ores are mined. A series of reactions commencing with the oxidation of the common mineral pyrite (FeS_2) is responsible, and certain autotrophic bacteria are closely involved (Lundgren *et al.*, 1972). The reactions may be summarised:

$$2FeS_2 + 2H_2O + 7O_2 \rightarrow 2FeSO_4 + 2H_2SO_4$$

$$2FeSO_4 + O_2 + H_2SO_4 \rightarrow Fe_2(SO_4)_3 + 2H_2O$$

$$Fe_2(SO_4)_3 + 6H_2O \rightarrow 2Fe(OH)_3 + 3H_2SO_4.$$

These reactions occur chemically, but only very slowly at acid pH values. The sequence of reactions may therefore be self-limiting. However, certain bacteria, especially *Thiobacillus thiooxidans* and *Ferrobacillus ferroxidans* possess enzymes which bring about these reactions rapidly. These bacteria are abundant in acid mine waters, especially since they are tolerant of low pH whereas many species commonly found in soil and water are not. Further, they are autotrophic, that is they require no substantial amount of organic matter for growth, and soon dominate the flora.

The effects of acid mine drainage are threefold. The low pH itself has adverse effects on the receiving water flora and fauna. It also promotes the solubilisation of heavy metals, which exert their own toxic effects. Thirdly, as the drainage water is diluted and the pH rises, ferric hydroxide precipitates and discolours the water, producing the effects of suspended particles. As the hydroxide settles, it forms a gelatinous layer over and within the substratum, causing both direct and indirect effects on the receiving water community. In general, the result is a marked reduction in species diversity and biomass in the affected areas (Koryak *et al.*, 1972; Letterman & Mitsch, 1978; Scullion & Edwards, 1980). The effect on the receiving water community is influenced by the alkalinity of the receiving waters or the presence of nearby alkaline discharges; where the acid is neutralised, the effects are due to the ferric hydroxide alone, rather than to the combined effect of hydroxide and low pH.

A phenomenon which has recently received widespread attention is that water bodies in some parts of the world, especially in North America and Northern Europe, appear to be becoming steadily more acidic. The reason, it is suggested, is that airborne pollutants from industrial areas are transported by prevailing winds to areas remote from their source, and precipitated in 'acid rain'. The subject has caused much controversy both within the scientific community and among the general public, particularly since the problem not only has grave implications for important economic interests, but has already given rise to differences between the governments of some countries. It is not yet clear, for example, whether the main cause is sulphur dioxide emissions from power stations, or motor vehicle exhausts. Some countries have already begun expensive programmes to limit SO_2 emissions, while others argue that the extent of acid rain is not controlled by the amount of SO_2 emitted but by chemical reactions in the atmosphere and that the reduction of SO_2 emissions would therefore not lead to a decrease in acid rain. It is also clear that certain forestry practices lead to acid runoff from the soil into surface waters, so that the source of the problem may lie not in distant industrial areas, but within the affected areas themselves. These controversies do not directly concern us, but it is appropriate to consider the extent to which natural waters are becoming acidified, whether or not this is a natural phenomenon, and what its biological consequences may be.

The consequences of acidification are undoubtedly serious. Haines (1981) provides a general review, and Baker & Schofield (1985) discuss the impact of acidification on North American fish populations. It is generally accepted that as pH decreases, both the diversity of species and the overall productivity of aquatic ecosystems declines. These phenomena can readily be observed through contemporary studies on waters of different pH values. Several mechanisms appear to be involved. Firstly, every species has its own zone of tolerance to pH. The pH range which permits survival may be rather wider than that which will permit successful reproduction. This is particularly well-documented for fish, both in laboratory experiments and through field observations (Baker & Schofield, 1985). Sutcliffe (1983) draws attention to the important influence of pH on the mechanisms of ionic regulation in aquatic species. Naturally-acid waters tend to be found in areas of base-poor geology and tend to be low in nutrients, dissolved ions and buffering capacity. Therefore they have low primary productivity, present unusual difficulty to animals and plants in maintaining ionic balance, and are particularly susceptible to extraneous inputs of acid. The distribution of invertebrates in relation to pH was investigated in a catchment in northern England (Sutcliffe & Carrick, 1973; Sutcliffe, 1983) and revealed a pattern which is widely repeated elsewhere (Haines, 1981). Phytoplankton, zooplankton and macrophytes behave similarly in relation to pH. Generally, the pattern is of reduced species diversity as pH declines, although in favourable circumstances acid-tolerant species may become locally abundant. Undoubtedly, an apparently small change in pH can bring about major changes in community structure. (It may be appropriate to recall that a reduction of 1 pH unit represents a tenfold increase in hydrogen ion concentration; and that a pH change from, say, 6.9 to 6.6 means that the hydrogen ion concentration has approximately doubled.)

In addition, low pH values strongly and progressively reduce the rate of decomposition of organic detritus, presumably through the effect of low pH on the fungal and bacterial organisms responsible for this process (Haines, 1981). Many aquatic ecosystems depend on the decomposition of allochthonous detritus (i.e. organic material from outside the system, such as fallen leaves) as the main source of energy for the animals in the system. Primary (photosynthetic) production of organic matter in acid waters is naturally low, being normally limited by the low availability of nutrients; it may also be itself inhibited by the susceptibility of phytoplankton and macrophyte species to low pH. The alternative source of energy for animals, organic detritus, is only available in the presence of microbial decomposers, since animals cannot digest plant material unaided. Therefore the overall productivity of the system will decline.

A third effect of acidification is to increase the threat of heavy metal toxicity. Aluminium, a metal which does not commonly cause serious problems of toxicity to aquatic life, has received particular attention. Burrows (1977) and Odonell *et al.* (1984) have summarised the role of aluminium as a toxic water pollutant. Although it is a very abundant metal, forming up to 7% of the earth's crust, it is highly reactive and readily forms stable compounds of very low solubility. In most natural waters which have been investigated, the dissolved aluminium concentration lies between 0.05 and 1.0 mg l^{-1}. Within the pH range of approximately 5.5 to 7.0, aluminium is practically non-toxic and is certainly harmless at the concentrations found in most waters. The chemistry of aluminium and its compounds at low pH is poorly understood, but it appears that as pH falls aluminium compounds become more soluble and that the proportion of free aluminium rises. Below pH 4, the toxic effects of free hydrogen ion are so severe that the presence of aluminium is probably of little significance. However, between pH 4 and 5.5, the toxicity of aluminium is high, reaching a maximum at around pH 5. At this pH, the level of aluminium naturally present in water is acutely toxic to fish. High levels of calcium appear to offer some protection, although acid waters with high calcium levels are unusual. Aluminium toxicity is likely, therefore, to be a major contributor to the effects of acidification. More rarely, aluminium can also cause toxicity in alkaline waters, as the solubility of aluminium compounds also increases with pH above 7.0. Aluminium compounds are widely used in the treatment of potable waters, and discharges from water-treatment works can have adverse consequences (Hunter *et al.*, 1980).

Finally, the question of whether acidification is a recent phenomenon, and whether it is a cyclic or an irreversible change, can only be answered indirectly. Accurate measurement of pH, particularly in weakly-buffered solutions like natural surface waters, has only been possible for about 30 years, and even within this timespan few water bodies have been studied systematically. It is also difficult to distinguish long-term trends from diurnal or seasonal pH variations, so the data available even from recent measurements of surface water pH are not altogether reliable. However, contemporary observations on the distribution of diatoms in relation to water pH show that the diatom community shows distinct qualitative changes according to the pH of the water. Analysis of the diatom community therefore indicates approximately the pH of the water. The characteristic siliceous frustules of diatoms accumulate at the bottom of lakes and decay only very slowly, so analysis of a core of lake sediment allows the changes in the diatom community over long periods of time to be determined. Thus, indirectly, any changes in the pH of the water may be inferred. Flower & Battarbee (1983) used this technique to estimate changes in the pH of two lakes in Scotland over several hundred years (Fig. 2.9), and

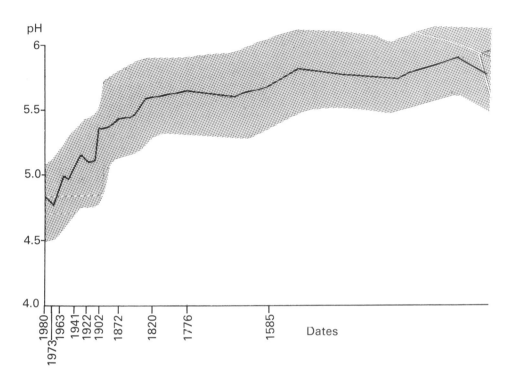

Fig. 2.9 — Changes in pH of a lake in Scotland with time, inferred from the composition of the diatom community in lake sediment cores at different depths (Flower & Battarbee, 1983). (Reproduced by permission from *Nature,* **305,** pp. 130–132; copyright © 1983 Macmillan Magazines Ltd).

their results are typical of many studies. It appears that pH of lake water declines slowly for natural reasons; rain is naturally acidic. However, the rate of decline of lake pH appears to have accelerated rapidly within the last 100 years or so, corresponding to the period during which Britain became heavily industrialised.

2.10 DETERGENTS

Synthetic detergents are an interesting group of pollutants because they were virtually unknown before 1945, yet within a few years became responsible for some spectacular water pollution problems which, unusually, came rapidly to the attention of the general public. The alkylbenzene sulphonate detergents (Fig. 2.10) rapidly replaced soap as domestic and industrial cleaning agents because of their cheapness and greater efficiency, and particularly because they did not cause precipitation of calcium salts in areas supplied with 'hard water'. Unfortunately they were not readily broken down by sewage treatment processes, giving rise to problems of toxicity to the receiving-water biota, and of foaming in watercourses and treatment works. In areas where industrial usage of detergents was pronounced (for example, in textile-processing industries) whole towns were frequently covered in detergent foam; in waste treatment works, a number of serious accidents occurred through, for example, operatives falling into sedimentation tanks which were concealed under a thick layer of foam. Consultation between regulatory authorities and the detergent manufacturers led to research which showed that modifications to the manufacturing process could produce *linear alkylate sulphonate* (LAS) detergents (Fig. 2.10),

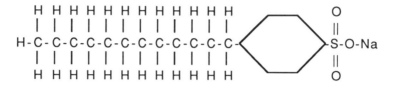

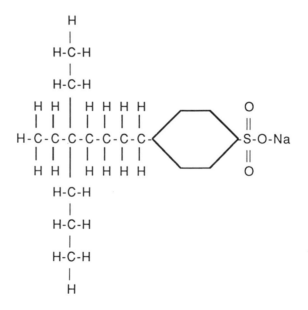

Fig. 2.10 — Molecular structure of (above) a typical 'soft' (biodegradable) linear alkylate sulphonate detergent and (below) a 'hard' alkylbenzene sulphonate detergent.

which were rapidly degraded in conventional waste treatment plants. From 1965 onwards, 'soft' or biodegradable detergents were introduced for domestic use, and although these are generally more toxic to aquatic organisms, their unbranched hydrocarbon chain is more readily broken down in treatment processes and in practice, toxicity and foaming problems had largely disappeared by the early 1970s.

Synthetic detergents remain significant causes of pollution in some circumstances however. Their toxicity to aquatic organisms is reviewed by Abel (1974). 'Soft' detergents are not suitable for use in certain industrial processes. Detergents are widely used as components of oil-dispersants, particularly in coastal and marine habitats, and are often more toxic to aquatic organisms than the oil itself (see Chapter 7). Finally, some components of detergent formulations exert adverse effects of their own. The best known example is the high level of phosphate found in many formulations (see section 2.4). Less widely appreciated are the adverse effects of boron, from perborate additives to detergent formulations, which can cause adverse effects on crops if contaminated surface waters are used for irrigation (Lester, 1975; see also Table 2.2).

3

Biological monitoring of water quality

3.1 THE CONCEPTUAL BASIS OF BIOLOGICAL MONITORING

We have by now seen ample evidence that the the levels of abundance and patterns of distribution of aquatic organisms may be affected by pollution of the water in which they live. Hellawell (1977, 1978) has summarised diagrammatically the changes which may occur in a community subject to pollution (Fig. 3.1). Which, if any, of these responses occur will depend upon the nature and severity of the pollution, and on the relative susceptibility of the species within the community to specific kinds of environmental alteration. It follows that given suitable techniques of sampling and data analysis, monitoring of the biological characteristics of waters might indicate the occurrence of ecologically significant environmental changes — including the incidence of pollution — which may otherwise be undetected.

Biological monitoring programmes are carried out for a variety of reasons. For example, most regulatory agencies, such as water authorities, routinely monitor the biota of the waters for which they are responsible. Typically, sampling stations are surveyed at intervals of between one and six months. Information is gathered on the presence and relative abundance of species, and may be used to derive numerical values, for example of species diversity or biotic index (see section 3.4), to facilitate spatial or temporal comparisons. Significant alterations from previously-established conditions may indicate the need for further investigation, and lead to action designed to preserve or improve the existing water quality. Sites known to be at special risk from existing pollution sources, and sites of special commercial or conservation value, for example salmon rivers, may receive special attention such as more frequent or more extensive surveillance. Proposed new developments, such as the siting of new effluent discharges, abstraction from or canalisation of water-courses, may be preceded by programmes of biological surveillance to establish 'baseline' biological conditions of the affected water. After implementation, the biota may continue to be monitored to determine the biological effects of the development and to assist in minimising them. Biological monitoring of receiving

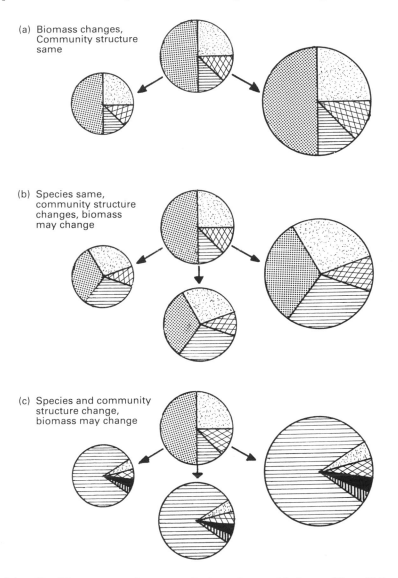

(a) Biomass changes,
    Community structure
    same

(b) Species same,
    community structure
    changes, biomass
    may change

(c) Species and community
    structure change,
    biomass may change

Fig. 3.1 — Possible responses of a community to environmental change. (From Hellawell, 1978).

waters may also, of course, be used to establish the effectiveness of effluent treatment processes and of the often expensive improvements to them which may be undertaken. Lemlin (1980) describes a good example of this approach. The effectiveness of a ten-year programme of improvement in the effluent treatment plant of an oil refinery was monitored by studying the distribution and abundance of saltmarsh flora in the area receiving the effluent. Since the cost of the alterations to the treatment plant was measurable in millions of pounds, confirmation that this cost was justified in terms of a measurable improvement in the receiving environment was of value to those responsible for the management of the refinery.

Living organisms can provide useful indications of the chemical quality of water. For example, chemical analysis of plant material may indicate contamination of the water with levels of heavy metals which may be potentially damaging in the long term, but which may not be otherwise detected (e.g. Whitton, 1980; Harding & Whitton, 1982; Whitton *et al.*, 1981; Say *et al.*, 1981; Whitton *et al.*, 1981). This approach has also been widely adopted in the marine environment, where levels of contaminants such as heavy metals or petroleum hydrocarbons in water may be below the detection limits of available analytical methods. Commonly, the levels of contaminants accumulated in the tissues of animals has been used to indicate the degree of chemical contamination of the environment. Sessile animals, particularly those which are distributed widely throughout the world such as the mussel *Mytilus edulis,* are particularly useful for this purpose (Goldberg *et al.*, 1978).

Biological monitoring, in conjunction with physical and chemical surveillance of water quality, generates information which is potentially useful in establishing water quality standards. To determine the level of a pollutant which is 'acceptable' (however that may be defined) is a frequently-recurring problem, and many toxicological and ecophysiological investigations are designed to provide information on this point. Complementary information may be provided by analysis of biological and chemical survey data, which may indicate directly the association between particular chemical conditions and the biological characteristics of the water.

Using modern techniques, chemical and physical analyses of water quality can be made with great precision. As will be seen, biological monitoring can be expensive of time and of human and physical resources. Its justification, however, is clear when some of the limitations of purely physical and chemical monitoring are considered. For example, a chemical analysis will only reveal the presence of the substance which the analysis is specifically designed to detect. The presence of a hitherto unsuspected pollutant would therefore not be detected by routine chemical surveillance, since it is uneconomic routinely to analyse water for large numbers of pollutants which might occur only very rarely. Biological surveillance, however, if properly carried out, will reveal the occurrence of ecologically-significant environmental changes and call attention to the need for further investigation.

Secondly, the concentration of pollutants in receiving waters fluctuates widely and rapidly. Even if samples are taken very frequently, it is highly probable that peak concentrations, which may be of greater biological significance than the 'background' levels, will be missed entirely (Fig. 3.2). Unlike a chemical analysis, a biological survey does not simply indicate the conditions prevailing at the instant of sampling. The organisms which live in the water respond to the totality of the environmental conditions which they have experienced throughout their lives. This is also important because pollutants usually occur in unpredictable mixtures rather than singly, and their action on aquatic organisms is subject to the modifying effects of environmental conditions (e.g. dissolved oxygen, temperature, water hardness) which themselves fluctuate and interact. The results of experimental studies are therefore difficult to apply to field situations (e.g. in terms of formulating water quality standards) in the absence of reliable information concerning the performance

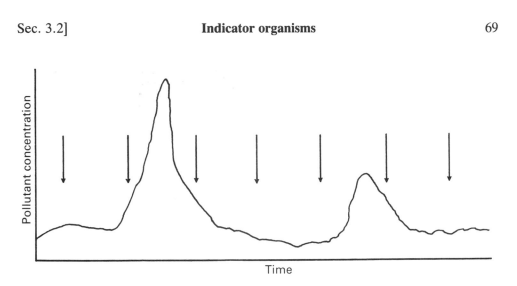

Fig. 3.2 — Chemical sampling, even at frequent intervals (arrows) can miss biologically-significant peak concentrations.

of species in polluted conditions in real rivers. Good biological monitoring programmes are thus complementary to toxicological and ecophysiological studies — each facilitates the interpretation and application of the results of the other.

3.2 INDICATOR ORGANISMS

It is fundamental to ecology that an organism cannot survive indefinitely in an environment that does not provide its physical, chemical and nutritional requirements. Thus the presence of a particular species, especially if it is reasonably abundant, indicates that its environmental requirements are being met. Its absence, however, does not *necessarily* indicate the converse — one species may, for example, be competitively excluded from a particular habitat by another. Nevertheless, within certain limitations the presence, absence or relative abundance of species may be used as indicators of environmental quality. Changes in the presence, absence or relative abundance of species, whether sudden or gradual, may therefore imply a corresponding change in environmental conditions. In its broadest sense, then, the term *indicator organism* can be used of any member of the fauna or flora of a habitat, and any species may be considered as a potential indicator organism.

Some organisms have such wide tolerances of different environmental conditions that their patterns of distribution or abundance are only slightly affected by quite wide variations in environmental quality. These species may tell us little about their environment which is not readily apparent by other means. The term *indicator organism* is therefore sometimes reserved to those species which have narrow and specific environmental tolerances, so that they will show a marked response to quite small changes in environmental quality. If the environmental factors which are most commonly limiting to the species concerned are known, its presence will be indicative of a specific environmental condition. Thus, nymphs of Plecoptera (stoneflies) are not found in waters where the dissolved oxygen concentration falls substantially below its saturation value for appreciable periods of time. Hence their presence indicates that the water is well oxygenated; and although their absence does not *necessarily* indicate the converse, their absence from waters wherein they might normally be expected at least suggests a possibility for further investigation. In practice, however, it is rare that enough is known of the ecology and physiology of a species for its presence or absence to be indicative of a specific environmental condition. There are, furthermore, very few organisms whose presence specifically indicates that the water is polluted. One exception may be, for example, the presence of coliform bacteria, indicating faecal contamination. Although several organisms are known frequently to be associated with polluting conditions — for example tubificid worms, and the 'sewage fungus' *Sphaerotilus natans* — they do occur widely in environments which are not polluted. In the present discussion, indicator organisms are considered to be 'those which, by their presence and abundance, provide some indication, either qualitatively or quantitatively or both, of the prevailing environmental conditions'. (Hellawell, 1978). The use of the term indicator organism in this sense should be distinguished from the use of the same term to apply to organisms whose tissue levels of contaminants are measured in order to infer the possible extent of chemical contamination of the environment. The term is widely used in both senses and it is important to distinguish between them.

Ideally all members of a community should be considered as potential indicators of water quality and included in biological monitoring programmes. In practice groups such as the bacteria, algae, protozoa and macroinvertebrates require such different sampling methods and taxonomic skills that most investigators choose only one such group. Normally this is found satisfactory except for the most extensive and detailed research programmes. The group most commonly employed as indicators is the macroinvertebrate fauna. They possess many of the characteristics required of indicator organisms, including:

(a) They are a diverse group in which some hundreds of common species from several different phyla are represented, so there is a reasonable expectation that at least some will respond to a given environmental change.
(b) They have relatively limited mobility and relatively long life cycles, so may sensibly be used for temporal and spatial analyses.
(c) Good identification keys are available for most macroinvertebrate groups, and it is possible to achieve a reasonable level of taxonomic competence quite quickly.
(d) Their high levels of abundance under favourable conditions facilitates quantitative analysis. Sampling techniques are fairly simple and well developed.

The disadvantages of macroinvertebrates as indicators, like their advantages, are common to some other groups. Seasonal variations in presence or abundance are a normal feature of many life cycles, so results need cautious interpretation. Many species drift passively downstream in substantial numbers, and may therefore be found in areas in which they cannot survive indefinitely. Finally, although it is relatively easy to obtain good qualitative samples of macroinvertebrates, reliable quantitative sampling is difficult owing to the physical characteristics of aquatic habitats and the complex horizontal and vertical distribution patterns of at least some species.

Although much of this chapter will be concerned specifically with invertebrates, many of the points discussed are of general relevance and it is undoubtedly true that other groups could, with advantage, be more widely used as indicators than is at present the case. The use of algae as indicators is discussed by Whitton (1979) and Shubert (1984). A scheme for the use of macrophytes as indicators of water quality has been proposed by Haslam (1982). The case for protozoa as indicators is put by Cairns (1974, 1979). The advantages and disadvantages of bacteria, protozoa, algae, macroinvertebrates, macrophytes and fish have been summarised by Hellawell (1978, 1986). More detailed discussions of several of these groups may be found in the volumes edited by Hart & Fuller (1974), James & Evison (1979), and Hellawell (1986).

3.3 SAMPLING METHODS

The validity of any ecological investigation depends crucially upon the sampling technique and strategy adopted at its outset. Many factors influence the design of a sampling programme. These include: the objectives of the investigation and the type of analysis to which the data are to be subjected; the physical characteristics of the habitat to be sampled; the characteristics of the organisms to be sampled — their size, habits, abundance and patterns of distribution; and constraints on the human, physical and financial resources available for the investigation. Frequently, then, a sampling programme is devised on the basis of a series of compromises between what the investigator would like to do, and what it is possible to do. In order to make sensible decisions about sampling — in particular, how to maximise the quantity and *quality* of the information obtained for a given sampling effort — it is necessary to understand something of the characteristics of the sampling techniques available. Good general descriptions of the various methods are given by Southwood (1978), Hynes (1970) and Hellawell (1978, 1986). Elliot (1977) and Hellawell (1978) discuss the statistical considerations underlying the choice of sampling strategies and the analysis of data. In the present discussion, some of the more commonly-used methods will be briefly described and their performance as qualitative and quantitative techniques examined.

Probably the most widely-used macroinvertebrate sampling method is the so-called 'kick sample'. A standard pond-net is held facing upstream, while the operator or a partner disturbs the substratum, preferably by hand rather than by actual kicking. Disturbed animals are washed into the net. Some operators, by sampling from a known area or for a standard period of time, attempt to introduce a quantitative element into the technique. However, it is very difficult in practice adequately to standardise either the area sampled or the duration of a sampling period, and almost impossible to standardise on the actual sampling effort, that is the degree of vigour or enthusiasm with which the sampling operation is carried out! Consequently large variation between samples and between operators is to be expected, and although it is possible to obtain good qualitative samples, the technique is rarely acceptable as a quantitative one.

The Surber sampler (Fig. 3.3) is designed to reduce some of the problems

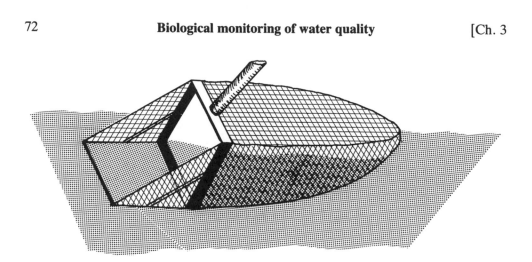

Fig. 3.3 — A Surber sampler. (From Hellawell, 1978.)

associated with operator variability and the standardisation of sampling effort and sampling area. A quadrat (usually 0.1 m^2) is attached to the frame of the collecting net in such a way that it can be placed on the substratum. The substratum within the quadrat is disturbed and animals are washed by the current into the net. Triangular vanes of netting at the sides of the sampler are designed to reduce the loss of material around the sides of the net. Box or cylinder samplers (Fig. 3.4) of various designs

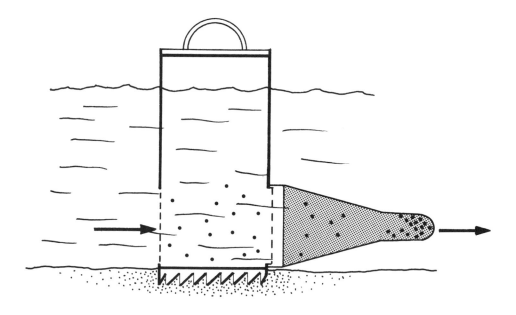

Fig. 3.4 — A cylinder sampler.

operate on a similar principle. Here the objective is to isolate as completely as possible the area to be sampled by pushing the sampler into the substratum. Both Surber and box/cylinder samplers can be used quantitatively, but both types can only be used in fairly shallow water.

For deeper water, grabs or dredges must be used. Both are available in a wide variety of designs (Figs. 3.5, 3.6). Grabs are the preferred method for fine substrata,

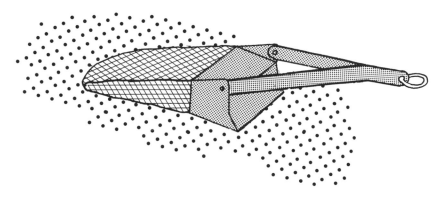

Fig. 3.5 — A dredge sampler (From Hellawell, 1978.)

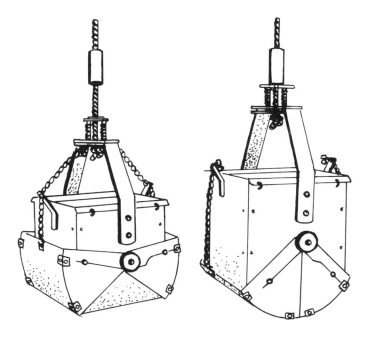

Fig. 3.6 — An Ekman grab in its open (left) and closed (right) positions.

but their efficiency is often low. Elliot & Drake (1981a) compared the performance of seven commonly-used designs. In one series of trials, the grabs were used to catch plastic beads (representing animals) from tanks containing known densities of beads in gravel of various sizes. The efficiencies of the grabs, in terms of the numbers caught as a percentage of the number available to be caught, are summarised in Table 3.1. The efficiency of all grabs was low when the modal particle size exceeded

Table 3.1 — Efficiencies of seven grabs at recovering animals from the surface and below the surface of two different substrata. Data from Elliot & Drake (1981a)

Grab	% Efficiency			
	Substratum 2–4 mm		Substratum 8–16 mm	
	Surface	3 cm below surface	Surface	3 cm below surface
Ponar	100	70	88	50
Weighted Ponar	100	70	100	50
Van-Veen	87	56	50	50
Birge–Ekman	73	37	30	7
Allan	51	36	25	7
Friedinger	59	7	30	7
Dietz–La Fond	22	26	25	20

16 mm, and some performed poorly under all conditions. Further, the grabs in general operated inefficiently when the 'animals' were buried 3 cm below the surface.

For sampling deep water with coarse substratum, dredges (Fig. 3.6) may be used. These are dragged along the bed collecting substratum and animals in the collecting net. Again, it is impossible in practice to standardise the area sampled and consequently to use dredges as reliable quantitative samplers. Elliot & Drake (1981b) assessed in field trials the efficiencies of four dredges as qualitative samplers. For each sample, a dredge was dragged along the river bed for five metres. The qualitative efficiency of each dredge was determined by expressing the number of taxa caught in five samples as a percentage of the total number of taxa caught at each site by all four dredges. Maximum values ranged from 76% for the most efficient to under 40% for the least efficient. These values were biased by the fact that the larger dredges caught more individuals; a more reliable indicator of the comparative performance of the dredges is that the least efficient design would need to be used 150 times to take a sample comparable with that obtained by using the most efficient dredge only five times.

A completely different sampling technique is to place artificial substrates in the water. These may take various forms (Fig. 3.7) for example baskets or trays filled

with stones, porcelain spheres, etc; 'multiplate' samplers, or bundles of plastic
material. Two designs which have been extensively evaluated in Britain are the slag

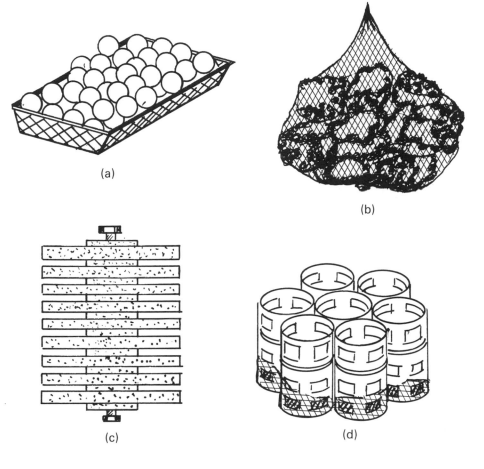

Fig. 3.7 — Some designs of artificial substrate samplers. (a) Wire tray or basket filled with
stones; (b) Slag bag; (c) Multiplate sampler; (d) SAUFU (standard aufwuchs unit).

bag, and the SAUFU (Standard Aufwuchs Unit) (HMSO, 1985). The slag bag
consists of standardised quantities of graded slag (as used in sewage filter-beds)
contained in a bag of plastic mesh. The SAUFU is constructed from a number of
units of synthetic plastic material, again originally designed for use in sewage filter
beds and similar processes, arranged in a circular configuration. After a period of
time they become colonised by plants and animals and can be recovered from the
water and the biota extracted. Sampling units may be standardised, for example by
using graded stones or completely artificial material. The advantages of standardisation, however, must be weighed against the possibility that some designs are likely

to be selective, that is the colonising fauna may be unrepresentative of the community sampled. Samplers may be embedded in the substratum, rested upon it, or suspended in mid-water. Typically it may take several weeks for samplers to be fully colonised, and frequently the numbers of taxa and numbers of individuals recovered reach a peak and then decline (Fig. 3.8). Thus it is important to standardise the colonisation period.

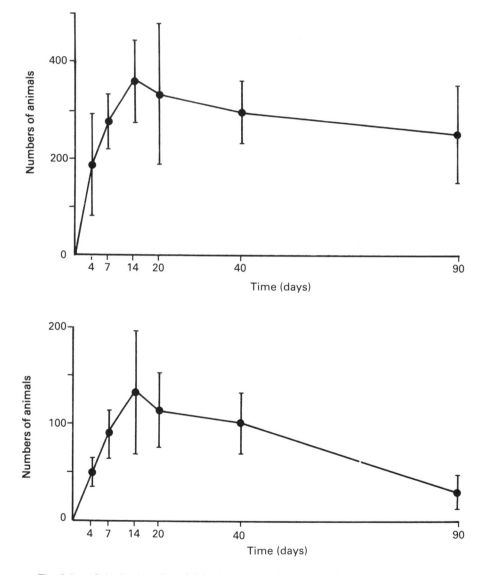

Fig. 3.8 — Colonisation of artificial substrate samplers exposed on a gravel river bed (upper graph) and a muddy river bed (lower graph) (From Hellawell, 1978, based on data given by Pearson & Jones, 1975).

3.3.1   Evaluating sampling techniques

An important characteristic of a qualitative sampling technique is the rate of taxon accretion, that is the relationship between sampling effort and the number of taxa caught. Obviously it is important to ensure that a qualitative sample contains a reasonably high proportion of the species actually present. To determine the extent of the sampling effort required to achieve this, a taxon accretion curve (Fig. 3.9) should be plotted. The correct procedure is to take a large number of samples, and

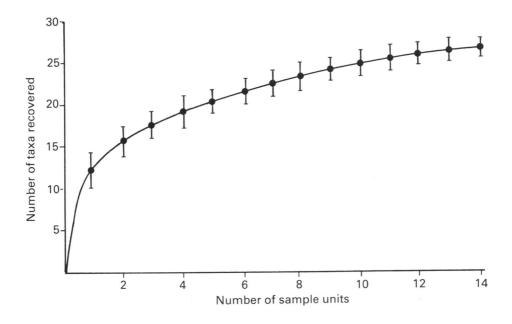

Fig. 3.9 — A taxon accretion curve.

determine the mean number of taxa recovered, with its confidence limits. Samples are then considered together in randomly-chosen pairs, threes, fours and so on. (A simple computer program can be used to calculate the values for all possible permutations of two, three, four etc.) A simpler method is to plot the cumulative number of taxa recovered against the number of samples taken, but this method can give misleading results owing to the possibility that the earlier samples may, by chance, contain abnormally high or abnormally low numbers of taxa. The resulting curve (Fig. 3.9), shows the point at which large increments in sampling effort yield few additional taxa, and the optimum sampling effort can thus be determined. As a guide, general experience shows that about five Surber samples, or 15–20 minutes of kick sampling, yields about 80% of the total number of taxa, though obviously this depends upon the competence with which the sampling is undertaken, and upon the characteristics of the habitat. Although it is rarely feasible to recover all the rarer taxa, it should be remembered that when making spatial or temporal comparisons between communities it is often the presence or absence of the rarer species that provides the most useful information.

Kick sampling is in practice the basis of most routine monitoring programmes, and an evaluation of the technique by Furse *et al.* (1981) yielded some interesting results. Four physically-different sites on a river were sampled by three experienced operators, each operator sampling each site for two periods of three minutes. The mean number of taxa caught per sample varied from 31.9 to 37.5 families, and from 48.3 to 58.3 species. Analysis of variance of the data showed that differences in the number of taxa per sample were significant with respect both to sites and to operators. The question then arises of whether the inter-operator differences are so large as to outweigh inter-site differences. If this were the case, kick sampling would be inadequate even as a qualitative technique since most monitoring programmes involve several different operators. To test this, the authors subjected their data to ordination and average-linkage cluster analyses. Six samples were taken at each site, and if inter-site differences were sufficiently large to outweigh inter-operator differences, the result of these analyses should be that each group of six samples should group together and be distinct from each other such group. (The principles of cluster analysis are described in section 3.4.)

Samples were first classified using similarity coefficients (see section 3.4) based on the presence or absence of *families*. This resulted in several 'misclassifications', that is samples which grouped with those from sites other than their own. The similarity coefficients were then recalculated to take into account the relative abundance of each family, rather than simply its presence or absence (the advantages of doing this are described in section 3.4). The resulting plots showed a marked reduction in the number of misclassifications. Perfect inter-site segregation could, however, be achieved by using data on the presence or absence of *species*. These results indicate that in addition to sampling technique, the methods of data analysis and the level of taxonomic skill employed are important contributors to the accuracy and validity of biological monitoring. An intrinsically weak sampling technique may yield biologically meaningful results if adequate taxonomic skills and analytical methods are brought to bear. In connection with the latter, however, caution must be exercised. It is a common weakness of many ecological investigations to use sophisticated analytical techniques on inherently poor data, as if mathematical manipulation could compensate for bad experimental design or inadequate sampling effort. A good test is whether the data analysis employed is biologically sensible. In the present case, if one site contains 99 individuals of taxon A and one of B, and another contains one of A and 99 of B, they are clearly biologically different. To work in terms of species presence and absence, therefore, is wasting valuable information — since both contain A and B, we might erroneously conclude that they are both the same. Thus the use of a more sophisticated coefficient of similarity is biologically justified.

Quantitative sampling is obviously more demanding than qualitative sampling. Estimates of relative population density are often useful in biological monitoring. It may be valuable to know, for example, that the abundance of a species has halved, or doubled, over time, or that the ratio of abundance of two species has altered. Provided that the data are interpreted cautiously, most sampling methods will yield some information on the relative abundance of the taxa recovered. However, Southwood (1978) discusses several factors other than the actual population density which influence the catch per unit sampling effort, and urges the adoption of a more careful approach to the interpretation of relative abundance data than is often shown.

The estimation of absolute population density is extremely difficult. The number of samples required to estimate population density is given approximately by the formula:

$$n = \left(\frac{st}{Dx}\right)^2$$

where n = the number of samples required, $\bar{x}$ and s are the mean, and its standard deviation, respectively of the number of individuals per sample caught in a pilot survey. t = the value of Student's t for the required level of confidence. D = the index of precision, that is the ratio of the standard error to the arithmetic mean expressed as a decimal.

Thus for example, if the mean number of individuals per sample is 10 and the standard deviation is 5, to obtain an estimate of the population density which lies within $\pm 10\%$ of the true value with 95% confidence,

$$n = \left(\frac{5 \times 2}{0.1 \times 10}\right)^2 = 100 \text{ samples.}$$

Unfortunately, the pattern of distribution of benthic invertebrates seems to be such that very large numbers of samples are required for reliable estimates of population density. Even to estimate population densities to within $\pm 20\%$ or $\pm 40\%$ of their true values may require several hundred samples (Hellawell 1977, 1978; Elliot & Drake, 1981a).

A further difficulty is that nearly all commonly-used sampling techniques are very superficial, in that only the top few centimetres of the substratum are sampled. That this can lead to serious errors is shown by the results of Coleman & Hynes (1970). They used artificial substrate samplers embedded 30 cm into the substratum and divided horizontally into four equal layers. Only 20% of the total catch was found in the top 7.5 cm and 26% was found between 22.5 and 30 cm deep. Clearly animals burrow deep within the substratum and only a small proportion of the population is recovered by normal sampling techniques. Radford & Hartland-Rowe (1971) compared a Surber sampler with an embedded sampler of 17.5 cm depth. Although the samplers produced similar qualitative results, the Surber took only 10% of the numbers and 53% of the biomass taken by an equivalent area of embedded sampler. Even allowing for the greater volume of substratum sampled by the embedded sampler their data suggest that smaller individuals, which may comprise a substantial proportion both numerically and in terms of biomass, are not adequately sampled by the usual techniques.

The limitations of sampling techniques are to a large extent a reflection of the characteristics both of the habitats under investigation and of the organisms which live in them. An appreciation of these limitations is essential at all stages of an investigation — in its design, in the analysis of the data, and in the formulation of its conclusions.

3.4 DATA ANALYSIS AND INTERPRETATION

The interpretation of biological survey data is essentially a series of comparisons — spatial, temporal, or both — and a variety of methods of data analysis are available to facilitate the process. Consider a single sampling station at a single point in time. Our sample or samples have yielded data in the form of a list of species, and perhaps an indication of their relative abundance. Assuming that the sample is reasonably representative of the community from which it was drawn, that is it has been obtained by an appropriate sampling method, we wish to infer from the sample data whether that community is affected by pollution. To know if pollution is affecting the community, we need to know whether that community is 'normal' or not. This can only be done by making temporal or spatial comparisons — is the community at this site the same as it was on some previous occasion? Or is it similar to physically comparable sites at different locations? Clearly in order to answer these questions we need to know the range of temporal and spatial variation in community chracteristics which may be considered 'normal'. To the extent that this knowledge is lacking, no matter how sophisticated our analytical techniques, a reasonable interpretation is impossible. Therefore a major need is for basic knowledge of the range of temporal and spatial variation in the community characteristics of *unpolluted* habitats.

In practice such information is available to varying degrees in particular cases, and the comparisons made will be determined by the extent of 'baseline' information available, and by the purposes of the survey. For example, comparisons may be made between different sites along the length of a river, or between samples taken from the same site at different times. Hellawell (1978, 1986) provides a very full discussion, with examples, of the methods which may be used to analyse and present the results of biological surveys. The results of such comparisons may be simply expressed as in Fig. 3.10, by plotting some characteristic of the community (in this case, numbers of species) against distance along the river. Such simple procedures may show a clear and comprehensible pattern, but obviously much of the information obtained from the survey is not being used. For example, there may be changes not only in the numbers of species present, but also in their relative abundance. Further, the species present at some sites may be completely different from those present at others. Such valuable biological information should not be wasted. One solution to this problem is to use one or more of the indices available (see below) as the variable which describes the communities. In Fig. 3.10 it can be seen that the Chandler Biotic Score more clearly demonstrates the biological impact of the effluents than does the number of species present at each station. However, it is not always the case that such derived values are more useful than the raw data, and it is always advisable to compare stations directly in terms, for example, of species numbers, numbers of individuals, and total biomass. Also, it is frequently informative simply to display the relative abundance of taxa (Figs 2.2, 3.11), which may clearly demonstrate differences between the various communities under comparison.

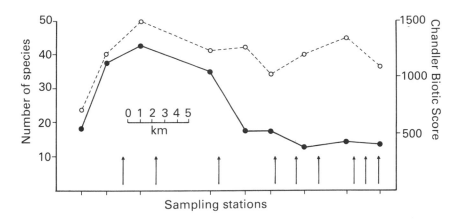

Fig. 3.10 — Number of invertebrate species (dotted line) and Chandler Biotic Score (solid line) at stations along the length of a polluted river. Arrows indicate inputs of pollution. Adapted from Hellawell (1978).

More sophisticated methods of analysis are frequently useful. Given that the data from a single site at a single point in time typically takes the form of a list of species and their relative abundance, and that comparisons between data sets may have to be made in both space and time, some means of condensing the data to manageable proportions is often essential. It is also important because although an experienced biologist may be capable of interpreting complex sets of ecological data by inspection, other specialists with whom he must co-operate (chemists, engineers, administrators) cannot be expected to do so. Therefore the raw data is frequently condensed by using it to compute one or more of several numerical indices or coefficients, of which four types can be distinguished — pollution indices, biotic indices, diversity indices, and similarity indices or coefficients. The applications of each of these will now be considered in turn.

3.4.1 Pollution indices
Pollution indices have reached a high degree of sophistication in many central and Eastern European countries (Sladecek, 1979). They are essentially developments of the descriptive Saprobien system of Kolkwitz and Marsson (1909) and are based on the fact that in rivers subject to organic pollution, communities downstream of the pollutant input show a regular and more or less predictable sequence of changes in the presence and abundance of indicator species, such as were described in Chapter 2 (see Fig. 2.2).

A typical pollution index is that of Pantle and Buck (1955). Organisms present in a sample are given an s-score and an h-score. The h-score is an index of relative abundance: organisms are scored 1, 3 or 5 depending upon whether they are rare, frequent or abundant respectively. The s-score relates to the saprobic zone of which the species are characteristic; a score of between 1 and 4 is assigned to each species, depending upon whether it is typically found in oligosaprobic, α-mesosaprobic, β-mesosaprobic or polysaprobic zones respectively. Central European workers have

Species

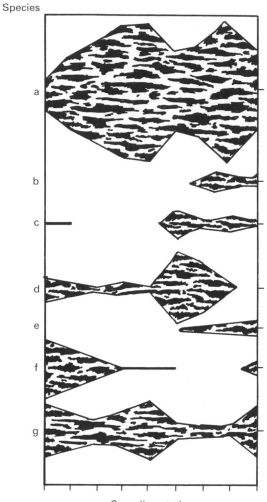

Sampling stations

Fig. 3.11 — The use of 'kite' diagrams to represent the abundance of species at different sites.
The width of the 'kites' represents the abundance of each species.

compiled extensive catalogues of the saprobic scores of invertebrate species, which are regularly updated as new information on distribution patterns is obtained. A good example is that of Wegl (1983). The saprobity index is given by

$$S = \frac{\Sigma sh}{\Sigma h}$$

and typically varies from less than 1.0 for unpolluted waters to 4.0 for heavily organically polluted waters. Hellawell (1978, 1986) and Washington (1984) give several other examples of pollution indices. Although they are widely used in continental Europe, they are less favoured in Britain and North America.

3.4.2 Diversity indices

Diversity indices were developed by theoretical ecologists who were interested in such questions as the relationship between stability and diversity in ecosystems. Washington (1984) lists no fewer than eighteen different diversity indices which have been used in water pollution studies (Table 3.2). Such a variety of diversity indices has arisen because ecosystem diversity is not easily defined, and therefore can be measured in several different ways, depending on how the concept is defined. Some simple examples will illustrate the point.

A simple measure of diversity is the number of species present in the community. However, it could be argued that this is an inadequate measure. For example, consider two communities of 1000 organisms. Community A contains one individual each of species a, b, and c, and 997 of d; community B contains 250 individuals each of a, b, c and d. Although each community contains four species, nearly all the individuals in A belong to the same species. Arguably, therefore, community B is a more diverse one than A. A numerical value derived both from the number of species in the community, and from the distribution of individuals between those species, is therefore a better index of community diversity. The application of diversity indices in water quality monitoring is discussed by Wilhm and Dorris (1968), Hellawell (1977, 1978, 1986), Kaesler *et al.* (1978) and Washington (1984). Diversity indices differ from one another in, for example, the relative weighting given to the number of species and to the distribution of individuals between species (the 'evenness' component). Some indices are based on specific assumptions of community structure, and many are not independent of sample size. Washington (1984) divides diversity indices into eight groups, according to the basis on which they are derived (Table 3.2). Having considered each in detail, he concluded that all were unsuitable for application to aquatic ecosystems except for Simpson's D (1949), Hurlbert's PIE (1971), indices based on the theory of runs (Cairns *et al.*, 1968; Keefe & Bergerson, 1976) and McIntosh's M (1967). The most widely used indices in practice are those based on information theory, such as the Shannon-Weaver index (Shannon and Weaver, 1949). However, as Washington (1984) points out, the biological relevance of such indices has been widely doubted in recent years.

Diversity indices seem to be particularly favoured by American workers, but are widely used throughout the world. A practical advantage of diversity indices is that it is not necessary to identify specimens. It is necessary only to recognise whether a particular specimen is of the same species, or of a different species, as previously-encountered individuals. Thus it is a useful method of analysis in circumstances where taxonomic skills are lacking, or where it is necessary to work with unfamiliar groups of organisms.

3.4.3 Biotic indices

In Britain, empirically-derived biotic indices of water quality are generally pre-ferred. These indices are based on two observed characteristics of the communities inhabiting polluted waters. Firstly, they generally contain fewer species than com-

Table 3.2 — Diversity indices which are widely used in aquatic ecosystem studies, arranged in eight groups according to their derivation. After Washington (1984)

(1) *Simpson's Index*
Simpson's D

where $D = \dfrac{\sum_{i=1}^{s} n_i (n_i - 1)}{n(n-1)}$ (1949)

(2) *Relatives of species number*
Kothé's species deficit

$\dfrac{A_1 - A_x}{A_1} \times 100$ (1962)

Odum's species per thousand individuals (Odum *et al.*, 1960)
(3) *Guesses by data fitting*

Gleason's index $D = \dfrac{S}{\ln N}$ (1922)

Margalef's index $D = \dfrac{S-1}{\ln N}$ (1958)

Menhinick's index $D = \dfrac{S}{\sqrt{N}}$ (1964)

(4) *Curve fitting approach*
Motomura's geometric series,

$y = Ac^{(x-1)}$ (Whittaker, 1965)

Fisher's α,

where $S_1 = \alpha \ln\left(1 + \dfrac{N}{\alpha}\right)$ (Fisher *et al.*, 1943)

The modified Yule' 'characteristic'

$\dfrac{M_1^2}{M_2 - M_1} = \dfrac{n^2}{\Sigma n(n-1)}$ (Williams, 1964)

Preston's log-normal 'a'

where $y = y_0 \exp(-aR)^2$ (1948)

(5) *Information theory*

Brillouins' $H = \dfrac{1}{N}\ln\dfrac{N!}{\prod_{i=1}^{s} N_i!}$ (1951)

Shannon's $H = -\sum_{i=1}^{s} \dfrac{n_i}{n} \ln \dfrac{n_i}{n}$ (Shannon & Weaver, 1949)

Evenness[†] $E = \dfrac{H'}{H'_{max}}$

Redundancy $R = \dfrac{H'_{max} - H'}{H'_{max} - H'_{min}}$ (Patten, 1962)

Table 3.2 (*continued*)

(6) Hurlbert's PIE $= \left(\dfrac{N}{N-1}\right)\left(1-\displaystyle\sum_{i=1}^{s} p_i^2\right)$ (1971)

(7) McIntosh's 'Ecological distance' relative

McIntosh's $M = \dfrac{n - \sqrt{\displaystyle\sum_{i=1}^{s} n_i^2}}{n - \sqrt{n}}$ (1967)

(8) *Theory of runs*
Cairns' SCI $=$
$\overline{DI_1} \times$ No. taxa   (Cairns *et al.*, 1968)

$\overline{DI_1} = \dfrac{\displaystyle\sum \dfrac{\text{no. runs}}{\text{no. specimens}}}{\substack{\text{no. times done to be} \\ \text{statistically significant}}}$

Keefe's $TU = 1 - \left(\dfrac{n}{n-1}\right)\left\{\displaystyle\sum_{i=1}^{K} p_i^2 - \dfrac{1}{n}\right\}$ (Keefe & Bergerson, 1977)

---

† See Washington (1984) for a discussion of Evenness.

List of terms
$S =$ the number of species in either a 'sample' or a 'population';
$K =$ number of taxa in either 'sample' or a 'population';
$N =$ the number of individuals in a population or community;
$N_1 =$ the number of individuals in species i of a population or community;
$n =$ the number of individuals in a sample from a population;
$n_1 =$ the number of individuals in a species i of a sample from a population;
$p_1 = n_1/n =$ the fraction of a sample of individuals belonging to species i;
$\Pi_1 = N_1/N =$ the fraction of a population of individuals belonging to species i.

Symbols of indices from the literature have been changed to conform to the above.

munities from comparable unpolluted waters; and secondly, as the degree of pollution increases species will tend to be selectively removed in order of their relative susceptibility to that form of pollution. The Trent Biotic Index (Woodiwiss, 1964) is most commonly used, and is reproduced in Table 3.3. To derive the index value, each specimen in the sample is identified to the level at which it can be assigned to one of the groups listed in the lower part of the table. It can be seen that in most cases fairly limited taxonomic expertise is required. When the number of groups in the sample is known, the table is entered at the appropriate column (e.g. if seven groups are present, enter the table at column 5). The line in the table at which the correct index value is given is determined by the presence of the highest-ranking indicator group (Column 1). Thus if seven groups are present, of which the highest-ranking member is *Gammarus*, the TBI value is V. If seven groups are present, but one is a species of Plecoptera, the TBI score is VII. Hellawell (1978, 1986) and Mason (1981) give worked examples showing the derivation of TBI values from raw data.

Table 3.3 — The Trent Biotic Index (Woodiwiss, 1964)

Key indicator groups	Diversity of fauna	Total number of groups (see Part 2) present				
		0–1	2–5	6–10	11–15	16 +
Column No: 1	2	3	4	5	6	7
		Biotic Index				
Plecoptera nymphs present	More than one species	—	VII	VIII	IX	X
	One species only	—	VI	VII	VII	IX
Ephemeroptera nymphs present	More than one species[a]	—	VI	VII	VIII	IX
	One species only[a]	—	V	VI	VII	VIII
Trichoptera larvae present	More than one species[b]	—	V	VI	VII	VIII
	One species only[b]	IV	IV	V	VI	VII
Gammarus present	All above species absent	III	IV	V	VI	VII
Asellus present	All above species absent	II	III	IV	V	VI
Tubificid worms and/or Red Chironomid larvae present	All above species	I	II	III	IV	—
All above types absent	Some organisms such as *Eristalis tenax* not requiring dissolved oxygen may be present	0	I	II	—	—

[a] *Baetis rhodani* excluded.

[b] *Baetis rhodani* (Ephem.) is counted in this section for the purpose of classification. The term 'Group used for purpose of the biotic index' means any one of the species included in the following list of organisms or sets of organisms.

Each known species of Plathyhelminthes (flatworms); Annelid worms (excluding genus *Nais*); genus *Nais* (worms); each known species of Hirudinea (leeches); each known species of Mollusca (snails); each known species of Crustacea (hog louse, shrimps); each known species of Plecoptera (stonefly); each known genus of Ephemeroptera (may fly) excluding *Baetis rhodani*; *Baetis rhodani* (may fly); each family of Trichoptera (caddis-fly); each species of Neuroptera larvae (alder fly); family Chironomidae (midge larvae except *Chironomus thummi*); *Chironomus thummi* (blood worms); family Simulidae (black-fly larvae); each known species of other fly larvae; each known species of Coleoptera (Beetles and beetle larvae); each known species of Hydracarina (water mites).

The Trent Biotic Index, originally devised for use in the River Trent catchment in England, is widely used throughout the UK, sometimes in modified form. It is a simple system requiring minimal sampling and taxonomic expertise, and the values it yields show a consistent relationship with some chemical variables such as BOD and permanganate values (Woodiwiss, 1964). However, it does not work well in all polluted waters. Since the ranking of key indicator species reflects their tolerance to organic pollution (or low dissolved oxygen levels), the system gives anomalous results when used with communities subject to some kinds of toxic pollution.

Consider, for example, the data shown in Table 3.4, which shows the composition of invertebrate samples taken in November 1982 from each of four stations on the River West Allen (see Chapter 1). Sites 7 and 6 + are heavily polluted with zinc; normally the water at these stations contains between 1 mg and 2 mg of zinc per litre. Site 4, approximately 5 km downstream, is moderately polluted, normally containing less than 0.5 mg l^{-1}, and site 6a is an unpolluted tributary which joins the West Allen between sites 7 and 6 + . The marked faunal impoverishment of the polluted sites 7 and 6 + caused by the zinc is evident from the raw data, but is not reflected in the Trent Biotic Index values (Table 3.4). Site 7, for example, has a score of 9, which might normally be taken to indicate good water quality. The reasons for this anomalously high index value are threefold, and may be used to illustrate some important defects in many biotic indices. Firstly, the index takes no account of abundance: one individual of a taxon contributes as much to the index value as one thousand individuals. Secondly, because the index is designed to assess the biological quality of waters subject to organic pollution, it gives a high score to stoneflies. Since stoneflies are not particularly sensitive to zinc, the result is unduly influenced by the presence of stoneflies. Thirdly, because there are only eleven possible Trent Biotic Index scores (0–10), the enormous variety of biological communities is made to conform to one of these eleven categories. A score of 7, for example, can be achieved in eight different ways. A sample consisting of two different stoneflies is scored identically with one containing no stoneflies or mayflies, one caddis larva and ten other groups. Consequently the index is often insensitive to major differences in community characteristics. In the present example, both sites 6a and 4 are assigned a score of 10, but inspection of the raw data indicates that they sustain different biological communities. For this reason, the system has recently been amended. In the modified version, the range of possible scores is increased (Table 3.5) so that scores between 0 and 15 are possible.

Some of these difficulties are overcome by the more sophisticated Biotic Score System of Chandler (1970). This system demands a higher level of taxonomic competence and also takes into account the abundance of the taxa present in the sample (Table 3.6). To determine the biotic score value, each taxon present is accorded a score according to its abundance in a standard sample and to its position in the ranking order of sensitivity to pollution. Note also that for pollution-tolerant taxa, increased abundance results in a *lower* score being assigned. The sum of the scores for the individual taxa gives the overall biotic score for the sample. Theoretically there is no upper limit to the score value, but in practice the scale is from 0–2000. The Chandler Biotic Score system is superior, in principle, to the simpler Trent Biotic Index. It has a wider range of possible values, it takes abundance into account, and it is less likely to be seriously influenced by the fortuitous presence of small numbers of unrepresentative taxa. Consequently it is able to distinguish more readily between different communities, and is more sensitive than the Trent Index to slight alterations in water quality. Table 3.4 shows Biotic Score values for the four West Allen sites. The difference between the polluted sites 7 and 6 + and the cleaner sites 6a and 4 is more clearly emphasised by the Chandler Biotic Score. Although the system fails to distinguish between sites 6a and 4, an experienced biologist would perhaps recognise that while a score of about 1000 is good for a small upland beck, it is not particularly high for a stream with physical characteristics of site 4.

Table 3.4 — Invertebrates collected by kick sampling from four sites on the zinc-polluted river West Allen and an unpolluted tributary (see text). Levels of dissolved zinc, and Trent Biotic Index and Chandler Biotic Score values are also shown

	Site			
	7	6a	6 +	4
Dissolved Zn, mg l^{-1}	1.28	0.04	0.76	0.38
Invertebrate taxa:				
Plecoptera				
Protonemoura meyeri	2	7	16	5
Isoperla grammatica	1	0	0	3
Amphinemoura sulcicollis	3	0	0	8
Leuctra fusca	0	3	2	24
Taeniopteryx nebulosa	0	0	0	2
Chloroperla tripunctata	0	0	0	1
Perla bipunctata	0	1	0	0
Perlodes microcephala	0	2	0	0
Brachyptera risi	0	0	3	0
Ephemeroptera				
Baetis rhodani	2	9	7	3
Ecdyonurus venosus	0	67	7	131
Rhithrogena semicolorata	0	7	5	13
Trichoptera				
Polycentropus flavomaculatus	1	0	2	1
Agapetus sp.	1	2	0	0
Rhyacophila dorsalis	0	1	0	2
Hydropsyche sp.	0	0	0	5
Philopotamus sp.	0	0	2	0
Coleoptera				
Limnius volckmari	1	16	0	0
Elmis aenea	0	1	0	0
Diptera				
Chironomidae	0	1	0	1
Simulidae	1	0	0	0
Dicranota	1	0	0	0
Tipula	1	10	1	0
Neuroptera				
Sialis lutaria	0	1	0	0
Mollusca				
Ancylastrum fluviatilis	0	11	0	0
Other taxa				
Gammarus pulex	1	35	0	2
Hydracarina	0	0	0	1
Hirudinea	0	0	0	1
Oligochaeta	0	0	0	1
Total no. of taxa	11	16	10	17
Total no. of individuals	15	174	46	204
Trent Biotic Index value	IX	X	VIII	X
Chandler Biotic Score value	588	1001	618	998

More recently a scheme known as the BMWP score (the initials stand for 'Biological Monitoring Working Party') has been undergoing evaluation in Britain (National Water Council, 1981). Like other schemes, it is largely derived empirically, and represents a compromise between ecological validity and practical constraints. For example the taxonomic requirement is limited to the need to identify specimens to the family level. In the BMWP scheme, specimens in a sample are identified to family and each specimen assigned a score according to Table 3.7. The BMWP score is the sum of the individual scores of the specimens. No account is taken of abundance, or of the fact that a family may be represented in the sample by more than one species. Unlike earlier schemes, the BMWP score system was not devised for any single catchment or geographical area; it is specifically intended to apply equally well to all areas of Britain, and therefore includes a number of taxonomic groups (e.g. Odonata and Hemiptera) which do not feature prominently in previous systems.

Several authors have attempted to compare the performance of pollution, diversity and biotic indices by using them to analyse sets of data (e.g. Nuttall & Purves, 1974; Balloch et al., 1976; Cook, 1976; Hellawell, 1978). Generally the Chandler Biotic Score has been found to be among the most satisfactory, and may be recommended as a good method for the routine assessment of water quality. Simpler indices tend to suffer to a greater degree from one or more of several disadvantages — insensitivity to major differences between communities, the production of anomalous or misleading results when non-organic pollution is involved, or undue bias in the result owing to the fortuitous presence or absence of small numbers of particular taxa. The BMWP index has not yet been fully evaluated but appears to have many useful features (Armitage et al., 1983). It has the advantages of taxonomic simplicity, and is applicable to a wider range of waters and geographical areas than other indices. For this reason it has been adopted for national survey purposes in Britain. Green (1984) concluded that the BMWP index could distinguish between the zinc-polluted West Allen and the unpolluted East Allen (see Chapter 1), although in the 1980 national survey it failed to do so (National Water Council, 1981). This was probably because the sampling procedures specified for the national survey were less rigorous than those applied in Green's more detailed study. It is prudent to conclude that minimal sampling effort combined with the more qualitative and taxonomically simple indices will inevitably lead to some false conclusions.

3.4.4 Similarity indices
Calculation of the degree of similarity between samples, like the calculation of diversity, can be done in various ways depending upon how similairity is defined. Thus there are a variety of similarity coefficients available for use.

One measure of similarity is the number of species common to both, as in Sorensen's (1948) coefficient:

$$S = \frac{2c}{a+b}$$

where a = the number of taxa in community a, b = the number of taxa in community b, and c = the number of taxa common to both.

Table 3.5 — The extended version of the Trent Biotic Index. Groups are defined as shown in Table 3.3

Biogeographical region: Midlands, England		Total number of groups present									
		0–1	2–5	6–10	11–15	16–20	21–25	26–30	30–35	36–40	41–45
		Biotic indices									
Plecoptera nymphs present	More than one species	—	7	8	9	10	11	12	13	14	15
	One species only	—	6	7	8	9	10	11	12	13	14
Ephemeroptera nymphs present	More than one species[a]	—	6	7	8	9	10	11	12	13	14
	One species only[a]	—	5	6	7	8	9	10	11	12	13
Trichoptera larvae present	More than one species[b]	—	5	6	7	8	9	10	11	12	13
	One species only[b]	4	4	5	6	7	8	9	10	11	12
Gammarus present	All above species absent	3	4	5	6	7	8	9	10	11	12
Asellus present	All above species absent	2	3	4	5	6	7	8	9	10	11
Tubificid worms and/or Red Chironomid larvae present	All above species absent	1	2	3	4	5	6	7	8	9	10
All above types absent	Some organisms such as *Eristalis tenax* not requiring dissolved oxygen may be present	0	1	2	—	—	—	—	—	—	—

[a] *Baetis rhodani* excluded.
[b] *Baetis rhodani* (Ephem.) is counted in this section for the purpose of classification.

Table 3.6 — The Chandler Biotic Score System (Chandler, 1970)

		Abundance in Standard Sample				
Groups present in sample		Present 1–2	Few 3–10	Common 11–50	Abundant 51–100	Very abundant 100 +
		Points scored				
Each species of	*Planaria alpina* Taenopterygidae, Perlidae, Periodidae, Isoperlidae, Chloroperlidae	90	94	98	99	100
Each species of	Leuctridae, Capniidae, Nemouridae (excluding *Amphinemoura*)	84	89	94	97	98
Each species of	Ephemeroptera (excluding *Baetis*)	79	84	90	94	97
Each species of	Cased caddis, Megaloptera	75	80	86	91	94
Each species of	*Ancylus*	70	75	82	87	91
—	*Rhyacophilia* (Trichoptera)	65	70	77	83	88
Genera	*Dicranota, Limnophora*	60	65	72	78	84
Genus	*Simulium*	56	61	67	73	75
Genera of	Coleoptera, Nematoda	51	55	61	66	72
—	*Amphinemoura* (Plecoptera)	47	50	54	58	63
—	*Baetis* (Ephemeroptera)	44	46	48	50	52
—	*Gammarus*	40	40	40	40	40
Each species of	Uncased caddis (exc. *Rhyacophila*)	38	36	35	33	31
Each species of	Tricladida (excluding *P. alpina*)	35	33	31	29	25
Genera of	Hydracarina	32	30	28	25	21
Each species of	Mollusca (excluding *Ancylus*	30	28	25	22	18
—	Chironomids (excl. *C. riparius*)	28	25	21	18	15
Each species of	*Glossiphonia*	26	23	20	16	13
Each species of	*Asellus*	25	22	18	14	10
Each species of	Leech (excl. *Glossiphonia, Haemopsis*	24	20	16	12	8
—	*Haemopsis*	23	19	15	10	7
—	*Tubifex* sp.	22	18	13	12	9
—	*Chironomus riparius*	21	17	12	7	4
—	*Nais*	20	16	10	6	2
Each species of	air breathing species	19	15	9	5	1
	No animal life			0		

If we have two samples, A and B, each containing two species, x and y, then according to Sorensen's coefficient they are 100% similar. However, if sample A contains 99 individuals of x and one of y, and sample B contains one of x and 99 of y, they are clearly biologically different, and Sorensen's coefficient is a misleading figure. Clearly there are advantages in using coefficients which take into account the relative abundance of species in the sample, and this can be done in various ways. Sneath and Sokal (1973) give a comprehensive list of similarity indices. Of the large number available, relatively few have been applied in aquatic studies and it is not yet

Table 3.7 — The BMWP score system

Allocation of biological scores

Families	Score
Siphlonuridae Heptageniidae Leptophlebiidae Ephermerellidae Potamanthidae Ephemeridae Taeniopterygidae Leuctridae Capniidae Perlodidae Perlidae Chloroperlidae Aphelocheiridae Phryganeidae Molannidae Baraeidae Odontoceridae Leptoceridae Goeridae Lepidostomatidae Brachycentridae Sericostomatidae	10
Astacidae Lestidae Agriidae Gomphidae Cordulegasteridae Aeshnidae Corduliidae Libellulidae Psychomyiidae Philopotamidae	8
Caenidae Nemouridae Rhyacophilidae Polycentropodidae Limnephilidae	7
Neritidae Viviparidae Ancylidae Hydroptilidae Unionidae Corophiidae Gammaridae Platycnemididae Coenagriidae	6
Mesovelidae Hydrometridae Gerridae Nepidae Naucoridae Notonectidae Pleidae Corixidae Haliplidae Hygrobiidae Dytiscidae Gyrinidae Hydrophilidae Clambidae Helodidae Dryopidae Elminthidae Chrysomelidae Curculionidae Hydropsychidae Tipulidae Simuliidae Planariidae Dendrocoelidae	5
Baetidae Sialidae Piscicolidae	4
Valvatidae Hydrobiidae Lymnaeidae Physidae Planorbidae Sphaeriidae Glossiphoniidae Hirudidae Erpobdellidae Asellidae	3
Chironomidae	2
Oligochaeta (whole class)	1

clear which indices are generally preferable (Washington, 1984). They have been more widely used by terrestrial ecologists.

Analysis of survey data by similarity indices offers some important advantages, and may in fact eventually be found superior to analysis by diversity indices. Comparisons may be made simultaneously in space and time, each site or sample being compared in turn with every other site or sample. The following simple example, based on unpublished data obtained by the author and Dr. G. Chabrzyk, illustrates the technique.

Samples of benthic invertebrates were taken from each of nine consecutive stretches of the River East Allen in northern England. Sites AA, B, D, F and H are riffles (stretches of turbulent water) and sites A, C, E and G are pools. Two similarity coefficients were used, Sorensen's (1948) and Raabe's (1952). The latter is one which takes into account the relative abundance of the species in the sample. Every possible pair of sites was compared, and the results displayed as a matrix (Fig. 3.12) Boxes in

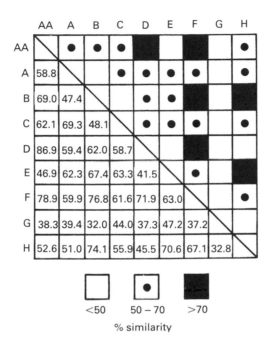

Fig. 3.12 — Matrix of Raabe's similarity coefficients for nine alternating pool and riffle stretches on the River East Allen.

the matrix can be shaded differentially, each pattern of shading corresponding to a different level of similarity. In some cases, visual inspection of the matrix allows any patterns to be detected. For example, in Fig. 3.12 it is immediately obvious that site G has a low similarity with any other site, which suggests that there is some unusual feature of this site which deserves further investigation. More detailed examination of the matrix reveals other interesting features.

There is no doubt that this approach is a very promising one. Its principal disadvantage is that the calculations involved are very tedious, particularly for large data sets. However, they can easily be programmed into a modern microcomputer of modest capacity. In fact, further treatment of the data greatly assists interpretation. The process of average-linkage cluster analysis (Williams, 1971; Pielou, 1984) allows a dendrogram to be constructed which displays visually the varying degrees of affinity between the sample stations (Fig. 3.13). Site G is now clearly seen to be dissimilar to all other sites, with a similarity to them of less than 40%. Two major groupings are now apparent — sites AA, D, B and F, with a similarity of at least 70%, and sites E, H, A and C with a similarity of just under 60%. All the sites on the

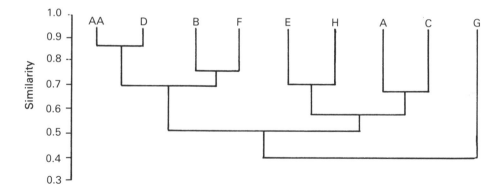

Fig. 3.13 — Dendrogram derived by average-linkage cluster analysis of the matrix shown in Fig.
3.12.

first group are riffles, and would be expected to have a high similarity. However,
another interesting result is that riffle site H has a high affinity with the pool sites E, A
and C. Thus the analysis has drawn attention to two stations, out of the nine sampled,
which have unusual or unexpected biological characteristics, and which perhaps
require further investigation.

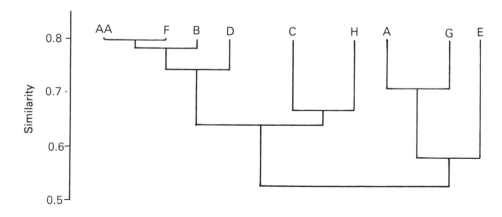

Fig. 3.14 — Dendrogram derived by average-linkage cluster analysis of a matrix of Sorensen's
similarity coefficients, based on the same raw data used to derive Fig. 3.13.

Fig. 3.14 shows the corresponding dendrogram based on the Sorensen coefficient
of similarity. Here the pattern is somewhat different. Again, riffle sites AA, B, D and
F form a group of high affinity, but the relationship of the 'anomalous' sites G and H
with the remaining sites is less obvious. The reason for this is clear from examination
of some of the raw data (Table 3.8). Sampling stations A and G are faunistically

Table 3.8 — Numbers of individuals of various taxa recovered from artificial substrate samples at two sites on the River East Allen. (See text and Figs 3.13 and 3.14)

Taxon	Nos. of individuals	
	Site A	Site G
Oligochaeta	6	51
Chironomidae	36	20
Limoniinae	32	4
Empididae	1	—
Limnius volckmari	1	1
Esolus parallelepipedus	1	1
Elmis aenea	1	—
Ancylastrum fluviatilis	6	—
Pisidium sp.	2	—
Polycentropus flavomaculatus	8	1
Baetis rhodani	—	3

poor, but have a number of taxa in common. However, site A is dominated by Chironomidae and Limoniinae in roughly equal numbers, whereas site G is dominated by oligochaetes with a much smaller number of chironomids, and other taxa present only in very small numbers. Thus with Sorensen's coefficient, but not with Raabe's, A and G appear very similar.

As argued above, the Sorensen coefficient is a purely qualitative one and arguably inferior to the Raabe one. The example is included to demonstrate the fact that the choice of coefficient can greatly influence the outcome of the analysis, and that sophisticated data analysis is not a substitute for sound biological reasoning, but an adjunct to it. Ideally, several coefficients should be used and the results compared before any definite conclusion is reached. Also, there are several different methods of cluster analysis (Williams, 1971; Pielou, 1984). Hellawell (1978) gives step-by-step descriptions of methods for constructing dendrograms.

A further level of sophistication is to estimate the degree of probability with which the sites are similar (or different). This is necessary because most samples do not include all the species present in the habitat. Thus two sites which are in fact identical may produce a similarity coefficient of 90%, 80%, 70%, 60%, or even less, depending upon the efficiency of sampling. At what level of similarity, therefore, do we conclude that two sites are significantly different? The question can be answered objectively by using, for example, Kendall's rank correlation coefficient (Kendall, 1962) in place of the more conventional indices of similarity. Examples are given by Hellawell (1978). This approach has been used in Figs 3.15 and 3.16, which relate to data on the invertebrate communities of 23 sampling stations on the Rivers East and West Allen (Green, 1984), and which indicate the application of cluster analysis in the study of a polluted river system. In these figures, stations numbered with Roman numerals are situated on the unpolluted River East Allen. Stations numbered with Arabic numerals are situated on the West Allen, which is heavily polluted with zinc in certain stretches. The similarity matrix (Fig. 3.15) immediately indicates certain

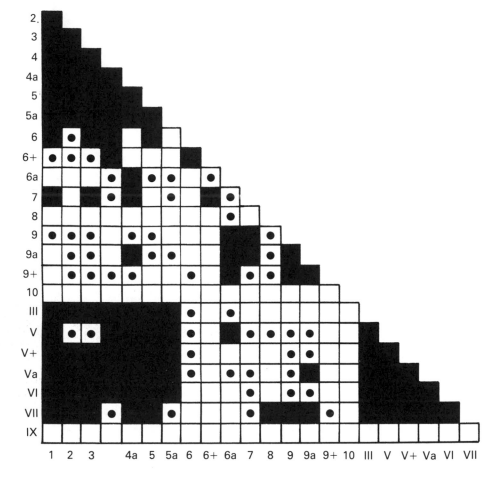

Fig. 3.15 — Matrix of Kendall's rank correlation coefficients for pairs of sites on the East and West Allen (Green, 1984). In this matrix, open squares indicate pairs of sites which are significantly different from each other.

stations which have low similarity to the remaining stations. Construction of the dendrogram (Fig. 3.16) allows a more detailed interpretation. In interpreting the dendrogram, we are particularly interested in those sampling stations which appear in an unexpected position. It must be remembered, however, that the appearance of a station in an unexpected position does not in itself indicate that the station is polluted.

At the level of Kendall's *tau* which indicates significant dissimilarity between the stations, seven groups, labelled A–G, are apparent (Fig. 3.16). Groups A and B each

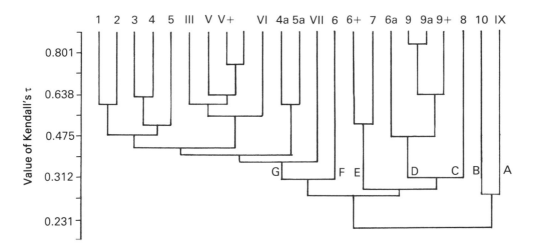

Fig. 3.16 — Dendrogram derived by average-linkage cluster analysis of the matrix shown in Fig. 3.15. See text for discussion. (Green, 1984.)

consist of only one station, IX and 10 respectively, which are thus both dissimilar from each other and from any other station. These sites are, in fact, the headwaters of the two rivers, where the physical conditions tend to promote unstable and highly variable communities. This grouping is not, therefore, unexpected. Group D contains four stations: 9, 9a and 9+ are all physically similar and adjacent to one another. The fourth station in the group, 6a, is a small tributary with similar physical properties to 9, 9a and 9+, so this grouping is again in accordance with expectations. Group A comprises 13 stations which include most of the downstream sites on both rivers. Note, however, that within this group two large sub-groups occur, comprising stations 1, 2, 3, 4 and 5 and stations III, V, Va, V+ and VI. Thus stations on the same river tend, as expected, to be more closely similar to one another than to stations on another river. The groupings which are anomalous are therefore C, E, and F, comprising stations 8, 6+ and 7, and 6 respectively. Chemical analysis of the water at these stations showed that it contained abnormally high levels of zinc (Abel & Green, 1981; Green, 1984). Thus the use of this technique allowed polluted sites to be identified from a large data matrix.

The analysis of data by similarity indices, although relatively recent in its application to aquatic ecosystems, is nevertheless a promising approach. It does not in itself allow a particular habitat to be diagnosed as 'polluted' but it does allow a large number of spatial and temporal comparisons to be made simultaneously. It draws attention to habitats which have unusual biological characteristics by comparison with the majority of habitats sampled, and thus allows further research to be directed where it will be most profitable.

3.4.5 Some general comments on indices
In view of the great variety of index methods available, and of the fact that relatively few have been fully evaluated, it is important that no single method be relied upon.

There is, unfortunately, a tendency to invest indices with a scientific validity which they do not necessarily possess, and it is necessary to emphasise some of their limitations.

First, it must be realised that the process of calculating an index results in the loss of most of the information represented by the raw data. The data contained in one sample typically consists of a list of species, tabulated to show the relative abundance of each. Such a table clearly contains many separate pieces of information. To reduce this to a single number, by calculating an index value, results in just one piece of information, the index value itself. Thus to rely solely on the index value is to reject most of the information which has been collected. That the consequences of this lead to error is easily demonstrated. Consider the Trent Biotic Index (Table 3.3). It has been seen that a score of VII can be obtained in at least eight different ways. So a sampling station which yields the same score on two different occasions may, in fact, have changed dramatically, and over-reliance on index values in the analysis of the data will actually obscure this fact. Similarly, a particular value of diversity index can be obtained from any of a large number of quite different samples. Two samples which have equal diversity index values may have quite a different composition, indeed they may be so different biologically that they have no species in common at all.

Secondly, comparison of index values, particularly diversity index values, from different sources (i.e. different laboratories) is extremely dangerous. Consider the widely used Shannon-Weaver index,

$$H = \sum_{i=1}^{s} P_i \log P_i$$

where P_i is the proportion of the total population belonging to the ith species and s is the total number of species in the sample. Some authors use $\log_2$, some use natural logarithms, and some use $\log_{10}$. Provided the same base is always used, no problem arises. However, most authors do not even state which base they have used, and therefore comparisons between results from different laboratories are meaningless.

A further difficulty arises through the use of different levels of taxonomic analysis. In theory every individual specimen is identified to the species. In practice this is rare, either because adequate keys do not exist or because individual laboratories lack suitably trained personnel. The extent of simple taxonomic error is, of course, unknown, but it is clear that different laboratories indulge to a varying, and usually undisclosed, extent in taxonomic 'lumping', which again renders inter-laboratory comparisons meaningless. Even within a single laboratory, during a single research programme, personnel changes can result in differing taxonomic standards being applied during a single investigation. Over-reliance on index methods, particularly those which require only identification to the family level, can cause major community changes to be completely overlooked. For example, the replacement of one species by another which belongs to the same family or genus is a biological change of potentially large significance, and precisely the sort of change which biological monitoring is intended to detect. Preoccupation with indices as methods of data analysis could cause an investigator completely to overlook even such a conspicuous alteration in community structure.

This does not mean that index values are useless in pollution studies. It does mean, however, that the index value is *not* the end-point of the data analysis, as many people seem to think. Index values themselves require biological interpretation. They are simply an aid to the analysis of the raw data, and should always be interpreted by referring back to the raw data *before* any conclusion is reached.

4

The toxicity of pollutants to aquatic organisms

Some examples of the toxicological data in the study of water pollution were illustrated in Chapter 1. There are many other circumstances in which the need to measure toxicity may arise. Many thousands of chemical substances are used for industrial, agricultural and domestic purposes, and their numbers increase annually. The toxicity of these chemicals, their by-products and degradation products to aquatic animals needs to be determined, since any compound manufactured and used in substantial quantities is likely to become a contaminant of watercourses. In the case of novel compounds or formulations, toxicity testing may precede large-scale manufacture and form part of the research into the feasibility of its commercial application. Toxicity tests may be incorporated into effluent monitoring schemes. Identification of the more toxic components of complex effluents may be a prerequisite for the development and improvement of effective treatment processes (e.g. Leach & Thakore, 1973). The measurement of toxicity is essential in the formulation of quality standards for receiving waters. Finally, compliance with a toxicity standard may be a legal requirement for consent to discharge an effluent. In this chapter we shall consider the basic principles of toxicology in relation to water pollutants and aquatic organisms, and the methods by which the toxicity and toxic effects of pollutants to aquatic life may be studied. The application of toxicological data in water pollution control, and their use in the formulation of water quality standards, will be considered in Chapter 6.

Toxic pollutants may exert their effects in several ways, depending upon the characteristics of the poison, of the receiving water, and of the biological community the water sustains. In extreme cases, animals may be killed by the poison. In some circumstances, poisons — insecticides, herbicides, molluscides, piscicides — are applied to water with the express purpose of killing some species and in the hope that others will be unaffected. Lower concentrations of poison may exert sublethal toxic effects. Some poisons appear to accumulate in the tissues of organisms during their lifetime, and exert toxic effects after prolonged exposure to concentrations which are barely measurable by chemical means. It is widely suspected that some of these may pass from prey to predator organisms and achieve high concentrations in species at the top of a food web. Many poisons are known to be mutagenic, teratogenic or carcinogenic, but the study of these phenomena in aquatic organisms is in its infancy.

From a biological point of view, any toxic effect is significant if it influences, or is likely to influence, the physiology or behaviour of the organism in such a way as to alter its capacity for growth, reproduction or mortality, or its pattern of dispersal, since these are the major determinants of the distribution and abundance of species. Species which are not directly affected by a pollutant may nevertheless be indirectly influenced. For example, if a predator is deprived of its normal prey by the action of a pollutant on the prey, it may itself be numerically reduced. Alternatively it may prey upon some other species, which itself may show a numerical response. Where two competing species are unequally affected by the pollutant, both may show a change in distribution and/or abundance. Thus the effects of toxic pollutants can only be fully understood with some knowledge of trophic, competitive and other interspecific relationships.

Since pollutants can exert such a variety of toxic effects at different levels of biological organisation, an enormously wide range of investigative methods have been employed in their study. There is an enormous literature on the subject of pollutant toxicity to aquatic organisms, especially fish and invertebrates. Much of it is, unfortunately, of limited or even doubtful value, for reasons that will be made clear. The whole field has been comprehensively reviewed in an excellent series of articles by Sprague (1969, 1970, 1971), and more recently by Abel (1988).

For the purposes of this discussion it is necessary at the outset clearly to define four basic terms which are widely misunderstood.

These are:

(a) *Lethal toxicity*: toxic action resulting in the death of the organism.
(b) *Sublethal toxicity*: toxic action resulting in effects in the organism other than its death.
(c) *Acute toxicity*: toxic action whose effects manifest themselves quickly (by convention, within a period of a few days).
(d) *Chronic toxicity*: toxic action whose effects manifest themselves over a longer period (by convention, within periods measurable in weeks or months rather than days).

A common misunderstanding among the ill-informed is to use the terms 'acute' and 'lethal' as if they were synonymous, and to do the same with the terms 'chronic' and 'sublethal'. This is clearly wrong. There are numerous examples of sublethal toxic effects which manifest themselves quickly, and of organisms which die only after prolonged exposure to a poison. Thus it is perfectly correct to talk of acute sublethal toxicity, acute lethal toxicity, chronic sublethal toxicity or chronic lethal toxicity and in many circumstances it is important to distinguish correctly between them.

4.1 LETHAL TOXICITY AND ITS MEASUREMENT

Approaches to measuring lethal toxicity vary in their complexity, in terms of the procedures employed, the apparatus required and the methods of collecting and processing the data produced. There are corresponding differences in the amount of information yielded, the degree of confidence which may be placed in the results, and the purposes for which those results may validly be used. There are many different reasons for carrying out toxicity tests, and it is important that the procedure chosen is appropriate to the purpose for which the results are required. Alabaster & Lloyd (1980) have discussed toxicity testing procedures in relation to their various applications. A useful practical guide to toxicity test methods has been published by HMSO (1983a). Since the results of toxicity tests may be significant in connection with pollution control legislation, national and international agencies have made some attempts to agree upon standard procedures (e.g. Ministry of Housing and Local Government, 1969; American Public Health Association, 1975; American Society for Testing and Materials, 1973; Maki & Duthie, 1978; Alabaster & Lloyd, 1980). Some of the simpler methods are widely used but provide very limited information, and their results need to be interpreted with more caution than is sometimes exercised. The reasons for this will become clear later in this chapter, but it is first necessary to understand the principles of more rigorous procedures. Some authors, notably Sprague (1969, 1970, 1973), have made specific recommendations on many aspects of toxicity test methodology, based on detailed consideration of the points discussed below.

4.1.1   The experimental conditions

The basic requirement is for groups of animals to be exposed to each of a series of concentrations of poison in suitable containers. A toxicity test does not differ from other types of experiment in that due consideration should be given to matters such as sample size, acclimation of animals to the experimental conditions, and maintenance of a constant environment. Sprague (1969) offers a detailed discussion of such points. Normally samples of at least ten are used, and the appropriate range of poison concentrations may be estimated on the basis of preliminary experiments so as to cause most of the animals to die over a period ranging from a few hours to a few days. However, it is often necessary to investigate the effects of lower concentrations such that the experiment may last for several weeks or more. The main practical difficulties encountered are in ensuring constant environmental conditions, maintaining the poison concentrations at their nominal levels, and minimising stress to the animals.

The toxicity of many poisons is greatly influenced by environmental conditions such as pH, temperature, water hardness and dissolved oxygen concentration. Clearly such variables should ideally be measured and controlled during any test, particularly since the presence of the animals themselves is likely to cause a gradual deterioration in the initial experimental conditions. This will arise due to utilisation of dissolved oxygen and excretion of carbon dioxide and other toxic metabolites such as ammonia. Poison concentrations may vary during the experiment due to absorption and metabolism by the animals, chemical and microbiological breakdown, or by evaporation. All these considerations suggest that the test containers should be fairly large and that the test solutions should be replaced regularly. Sprague (1969) recommends a minimum test volume of two to three litres per gram of animal tissue, to be renewed at least once daily. The preferred arrangement, however is to construct a 'continuous flow' apparatus in which the test solutions are automatically replenished, usually achieving a total replacement of solution every six to eight hours. The replacement rate may be greater or less depending upon the size of the test containers, the size, activity level and metabolic rate of the animals, and the volatility or degradability of the poison under test. Several designs of such apparatus have been published (e.g. Mount & Brungs, 1967; Abram, 1973). In practice, most investigators use apparatus of their own design and construction, and of varying degrees of complexity. Sprague (1969) and Abel (1988) refer to several alternative methods.

4.1.2 Data collection and analysis

The raw data from a toxicity test take the form of a record of increasing mortality in each test container as time passes. Ideally, the survival time of each individual in the experiment should be recorded accurately. In practice, however, it is not necessary to observe the experiment continuously throughout its duration. Sprague (1973) recommends observations at 0.25, 0.5, 0.75, 1, 2, 4, 8, 14±2, 24, and 33±3, hours, and thereafter at daily intervals.

The next step is to estimate the median survival time (LT50) of each group of animals, that is the time required for half the animals to die. The most widely-used technique is the rapid graphical method of Litchfield (1949), which is derived from a procedure developed by Bliss (1935, 1937). For each group of animals, a graph is plotted of cumulative percentage mortality against elapsed time. For reasons which will be discussed later, the mortality data are transformed to probits (or plotted on a probability scale) and the time values are transformed to log times. The straight line of best fit is then drawn through each set of points. This line may be computed, but in practice lines fitted by eye are often satisfactory. The result is a set of 'probit lines' or time-mortality curves as shown in Fig. 4.1. Values of median survival time (LT50)

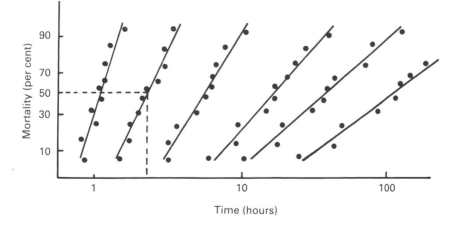

Fig. 4.1 — Probit lines resulting from plots of cumulative percentage mortality in each test tank against time, using logarithmic–probability graph paper. The dotted lines show how median survival times are read from each line. Lines on the left are for the higher poison concentrations.

can now be read off each line as shown. The time for each percentage responses, for example LT10, LT90, can also be read off if required. Normally values for LT16, LT50 and LT84 are required. To estimate confidence limits, the *slope function, S,* for each probit line is determined by

$$S = \frac{\dfrac{LT50}{LT16} + \dfrac{LT84}{LT50}}{2} \ . \ .$$

Using the values of S and N (where N = the number of animals in the test tank), the factor f is computed:

$$f = \text{antilog}\left(\frac{1.96 \log S}{\sqrt{N}}\right) = S^{1.96/\sqrt{N}}.$$

Litchfield (1949) provides a nomograph for the computation of f from values of S and N. The procedure is slightly more complex if not all the animals die during the experiment, or if a 'split' probit line (see below) is obtained. The upper and lower confidence limits of the LT50 are given by LT50×f and LT50÷f respectively. LT50 values and their confidence limits for each group of animals, can be plotted against poison concentration to give a toxicity curve (Fig. 4.2). It is customary to use logarithmic scales on both axes, since the ranges of survival times and poison concentrations to be displayed frequently span several orders of magnitude. This transformation does not, of course, alter the shape of the curve.

An alternative method of plotting the raw data is widely used. This method produces essentially the same toxicity curve but it is arrived at by a different route. For each observation time, a graph is plotted of percentage mortality (transformed to probits) against poison concentration (transformed to log concentration). Instead of a series of survival time-mortality curves (Fig. 4.1) we now have a series of concentration-mortality curves (Fig. 4.3). For each observation time, the median lethal concentration or LC50 (sometimes termed median tolerance limit, TL_M) can be read off. This value is defined as the concentration causing half the animals to die *within a specified period of time.* Its confidence limits can be estimated by a procedure analogous to that described above (Litchfield & Wilcoxon, 1949), based on the original method of Bliss (1935). The resulting toxicity curve (Fig. 4.4) of LC50 against observation time differs only in that confidence limits are expressed in terms of concentration rather than time. For certain applications this approach has some advantages; for example, where the test forms part of a research programme designed to establish water quality standards, it is obviously preferable to estimate errors associated with lethal concentrations rather than with survival times. However, the computation of the results is more complex since, among other difficulties, the number of points on each probit line is small. Unless the chosen range of concentrations is narrow, most groups of animals in the test will show either zero or

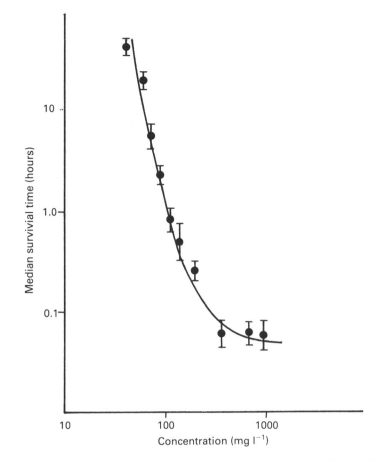

Fig. 4.2 — Toxicity curve relating median survival time to poison concentration; rainbow trout (*Salmo gairdneri*) exposed to sodium lauryl sulphate (Abel, 1978). Vertical bars represent 95% confidence limits.

100% mortality at any single observation time. Thus the lines of best fit must be calculated rather than fitted by eye (Litchfield & Wilcoxon, 1949).

We may now consider the interpretation of lethal toxicity test results:

4.1.3 Probit lines

The theoretical basis of the log-probit transformation of mortality data is given by Bliss (1935, 1937) and Finney (1971). This technique of data analysis is widely used in many types of toxicological and pharmacological research. Hewlett & Plackett (1979) provide a lucid and succinct account of the general theory and practice of probit analysis, and Sprague (1969) discusses its application to toxicity tests involving aquatic animals. In practical terms, the purpose of the transformation is to allow the estimation of median survival times and/or median lethal concentrations, and their confidence limits, from a relatively small sample. Other transformations have occasionally been used (Sprague, 1969) and some have been discussed in detail by

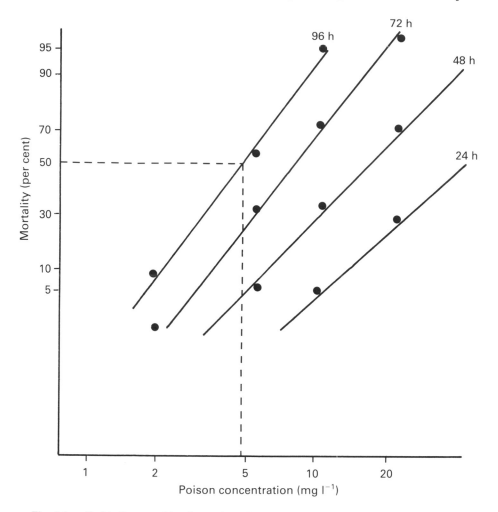

Fig. 4.3 — Probit lines resulting from plots of percentage mortality at specified observation times (probability scale) against poison concentration (log scale). The dotted line shows how the 96 h LC50 value may be read off.

Stephan (1977), who argues that the log-probit transformation is not necessarily the best. However it is by far the most widely used, and in practice all appropriate transformations appear to produce very similar results.

    Because the probit lines are an essential intermediate stage in the construction of the final toxicity curve, it is often overlooked that they can provide other information of toxicological interest. Useful indications are provided by examination of probit lines for irregularities of slope or the presence of inflections. 'Split' probit lines (Fig. 4.5) occasionally occur, which indicates a heterogeneity of the population from which the sample was drawn (Hewlett & Plackett, 1979). Such heterogeneity may be due to intra-specific variability in susceptibility to the poison between sexes or age classes, or may indicate the development of resistant strains within the population.

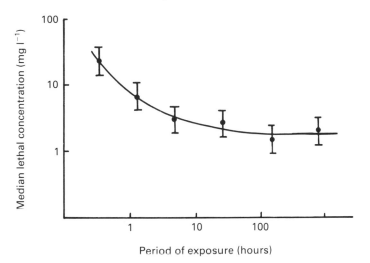

Fig. 4.4 — A toxicity curve in which median lethal concentrations and their confidence limits have been plotted against time. See Fig. 4.15 for other examples.

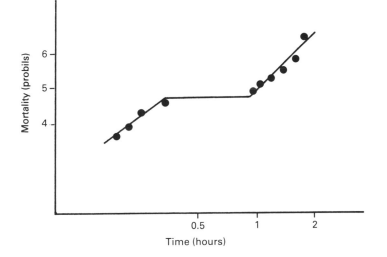

Fig. 4.5 — A 'split' probit line from an experiment to determine the lethal toxicity of the detergent sodium lauryl sulphate to rainbow trout, *Salmo gairdneri*. The discontinuity indicates a heterogeneity in the population of test animals, in this case due to the existence of two separate mechanisms of toxic action (Abel, 1978).

Another possible explanation is that the poison may have two (or more) mechanisms of action, and that animals which are resistant to one mechanism may subsequently succumb to another. Tyler (1965) concluded on the basis of split probit lines that high temperature had three separate mechanisms of lethal action in fish. Confirmation that such an interpretation can be correct is given by studies on the toxicity and toxic effects of an anionic detergent (Abel, 1976, 1978). A concentration-dependent change in the mode of lethal action of the detergent was found to be associated with the occurrence of split probit lines in toxicity tests.

Where the heterogeneity in susceptibility to the poison is not pronounced, it may be manifested by an alteration in the slope of the probit lines rather than in the occurrence of inflections (Hewlett & Plackett, 1979). Simple tests for significant differences in the slope of probit lines are available (e.g. Litchfield, 1949) and have been used to demonstrate the occurrence of more than one mode of lethal action in fish exposed to toxic pollutants (Burton et al., 1972; Abel, 1978).

Ball (1967a) showed that the slope of the probit lines was an important datum in comparing the relative susceptibilities of species to poisons. Ammonia was equally toxic, in terms of the 5-day LC50 value, to roach (Rutilus rutilus) and rudd (Scardinius erythrophthalmus). However, there was a considerable difference in the slope of the concentration-mortality curves. In practical terms, this may be important in the application of the toxicological data to water quality standards. For example, an ammonia concentration equivalent to two-thirds of the 5-day LC50 may be expected to kill less than 1% of rudd, but 16% of roach.

These examples illustrate the potential importance of a full analysis of probit lines in the investigation of lethal toxicity. Such detailed consideration is rarely given, and it would appear that much useful information from toxicity tests may thereby be lost. Where irregularities of the kind described above are found to occur, they may indicate promising lines of further investigation.

4.1.4 Toxicity curves

The toxicity curve describes the empirical relationship between poison concentration and survival time of the animals. Some representative curves are shown in Figs 4.2–4.9. There have been several attempts at mathematical descriptions of toxicity curves (reviewed by Sprague, 1969), but the validity and usefulness of this is questionable (Brown, 1973), as is speculation concerning the shape of the curve. As Brown (1973) points out, for any poison there will be a concentration so low that it will never cause the death of half the animals. Thus the curve will become asymptotic to the time axis. The concentrations at which this occurs may be termed the *threshold median lethal concentration* or threshold LC50. (A synonymous term is *incipient lethal level* or ILL.) Further, even at very high concentrations death will not be instantaneous, but will take a finite period of time to occur. Thus the curve will become asymptotic to the concentration axis. Between these two asymptotes, the curve generally will show a decrease in survival time as concentration increases. Whether the toxicity curve is a straight line or a curve is therefore of no significance. Where a straight line is obtained, it may be considered as a segment of a larger curve. This point is illustrated by Brown (1973), and also in Fig. 4.6. Here, toxicity curves for several species of fish exposed to some synthetic detergents have been redrawn from the originals to the same scale. Clearly, no significance can be attached to the

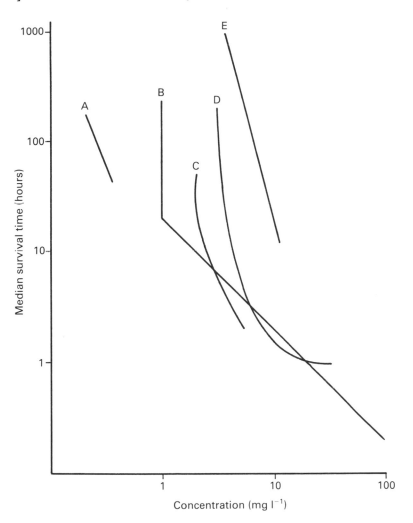

Fig. 4.6 — Some toxicity curves for various species of fish exposed to different detergents, redrawn from the originals to the same scale. A — *Salmo gairdenri*, linear alkylate sulphonate (Brown *et al.*, 1968). B — *Salmo salar*, polyoxyethylene lauryl ether (Wildish, 1972). C — *Lepomis macrochirus*, linear alkylate sulphonate (Hokanson & Smith, 1971). D — *Gadus morrhua*, alkylbenzene sulphonate (Swedmark *et al.*, 1971). E — *Salmo gairdneri*, alkybenzene sulphonate (Herbert *et al.*, 1957).

fact that, for example, curves A and E are rectilinear while C and D are curvilinear. Most of these curves, and A and E in particular, cover only part of the range over which survival time changes with poison concentration. More extensive testing over a wide range of concentration is required to establish the true shape of the curves.

The most important feature of a toxicity curve is the indication it gives of the threshold median lethal concentration, and wherever possible tests should be continued until a lethal threshold is apparent. Fig. 4.7 shows clearly the importance

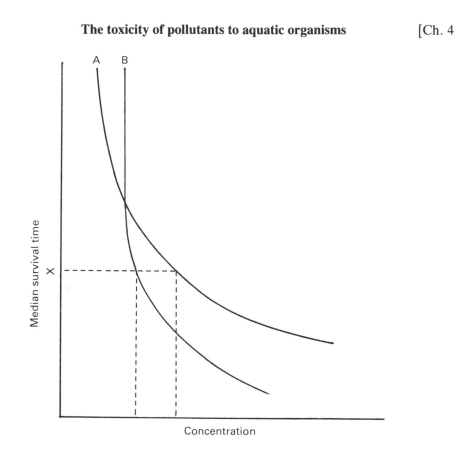

Fig. 4.7 — Two hypothetical toxicity curves. Substance A is clearly more toxic than substance B, since the lethal threshold concentration of A is smaller. However, had the test been discontinued at time X, B would have appeared more toxic than A.

of determining lethal threshold concentrations. In this diagram, two toxicity curves are shown. These may represent two poisons tested against the same species, or one poison tested against two different species, or one poison tested against one species under different environmental conditions. Clearly A is more toxic than B, since the lethal threshold concentration of A is smaller than that of B. If, however, the experiment had been terminated arbitrarily at time X, as shown on the diagram, it would be erroneously concluded that B is more toxic than A. Therefore unless complete toxicity curves are obtained, and lethal thresholds established, the results of any test must be interpreted with caution. This is particularly important where comparisons between species, poisons or environmental conditions are involved. For example, Ball (1967a) found that in tests lasting one day, trout were more sensitive to ammonia than coarse fish. However, when the tests were continued for up to five days, all fish were found to be equally sensitive, the difference between trout and coarse fish being simply that the latter were more slow to react.

Apart from being the only proper basis for comparative studies of lethal toxicity, lethal thresholds are also useful, in conjunction with other information, in setting water quality standards (see Chapter 6). A major criticism of many investigations of lethal toxicity is that experiments are not continued long enough. Sprague (1969) examined 375 published measurements of lethal toxicity and found that only 211 of these showed a lethal threshold within four days. In 122 cases the time required to show a lethal threshold was between four and seven days, and 42 cases required longer than this. Although the results of shorter experiments are by no means invalid, it is nevertheless clear that unless lethal thresholds are clearly established the interpretation which may validly be put on the results is strictly limited. Further, useful toxicological information may be lost. Ball (1967b) reported an example of the advantages of continuing tests as long as is economically or practically feasible (Fig. 4.8). The toxicity curve for rainbow trout exposed to cadmium was linear over

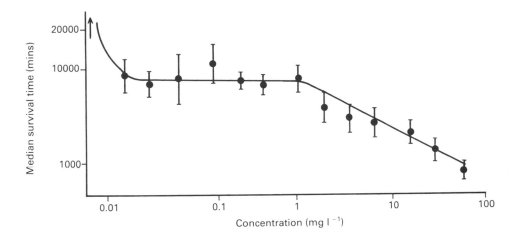

Fig. 4.8 — Toxicity curve for rainbow trout (*Salmo gairdneri*) exposed to cadmium (Ball, 1967b).

a concentration range between 1 and 64 mg l^{-1} and a time period of about six days. Continuing the test for 14 days revealed that cadmium continued to act lethally down to 0.01 mg l^{-1} and that the threshold concentration may lie as low as 0.008 mg l^{-1}. Fish exposed to concentrations between 0.01 and 1.0 mg l^{-1} continued to die throughout the latter part of the experiment, and between these two concentrations survival time did not increase as the cadmium concentration decreased. These results demonstrated that cadmium was a very slow-acting poison which was considerably more toxic than shorter tests had previously indicated. Toxicity curves for five fish species exposed to cadmium have been compared (Abel & Papoutsoglou, 1986) and show some interesting features (Fig. 4.9).

4.1.5 Some alternative methods for measuring lethal toxicity
As indicated earlier, there are many different reasons for measuring the toxicity of pollutants to aquatic organisms, and it is important to use a method which is appropriate to the purpose for which the results are required. The procedures

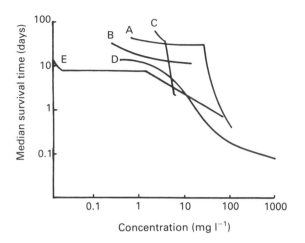

Fig. 4.9 — Toxicity curves for several fish species exposed to cadmium, redrawn to the same scale. Note that none of the curves shows a clear threshold concentration, despite the long duration of most of the tests. Inflections in the curves, and segments of curve over which mortality is apparently unrelated to poison concentration, commonly occur. Finally, the curves for different species frequently intersect: which species appears most, or least, sensitive depends upon the duration of the toxicity test. A — *Tilapia aurea* (Abel & Papoutsoglou, 1986). B — *Cyprinus carpio* (Abel & Papoutsoglou, 1986). C — *Noemacheilus barbatulatus* (Solbé & Flook, 1975). D — *Gasterosteus aculeatus* (Pascoe & Cram, 1977). E — *Salmo gairdneri* (Ball, 1967b).

described above are unnecessarily elaborate for some purposes, and similarly there are circumstances in which these conventional methods provide inadequate information. We will therefore consider alternative methods, on the one hand simpler, and on the other more complicated, than those discussed already.

For many purposes — for example rapid screening of novel compounds, routine monitoring of effluents, or for rapid experiments preliminary to a fuller investigation — the full procedure described above is not technically or economically feasible or, perhaps, necessary. Several simplified procedures have been proposed of which the most widely used is probably the American Public Health Association Standard Method (American Public Health Association, 1975). Fairly simple apparatus may be used, and observations of mortalities are made every 24 hours up to 96 hours. In its simplest form, the test provides a rough estimate of the 96 h LC50 (concentration required to kill half the animals in 96 hours) by graphical interpolation. The term TL_M (median tolerance limit) or TL_{50} (tolerance limit for 50 per cent of the animals) is used and is synonymous with LC50. Two points, representing survival at two successive concentrations that were lethal at 96 hours to more than half and less than half of the animals, are plotted on a semilogarithmic graph with poison concentration on the log scale. A straight line is drawn between the points, and the concentration at which this line crosses the 50% survival line is the TL_{50} value (Fig. 4.10). Sprague (1969) provided a critique of this test in its simplest form. Although it is acceptable for its purpose, that is routine screening and monitoring, the arbitrary termination at 96 hours subjects it to the limitations on the value of its results as described above.

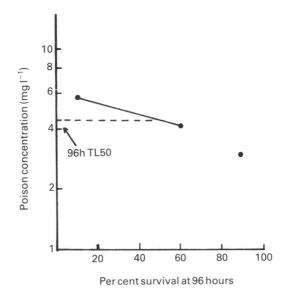

Fig. 4.10 — Estimation of median tolerance limit (TL_{50}) by the graphical interpolation method (American Public Health Association, 1975). The method gives only a very approximate estimate, with no confidence limits.

Further, in its simplest form the test does not gather all the information it could yield. More frequent observations of mortality, combined with more sophisticated techniques of data analysis and longer test duration, provide far more valuable results at little extra cost or effort. The protocol currently in use does, in fact, include reference to these points.

Tests for legal purposes must be simple yet reproducible, and prescribed sufficiently rigorously for their results to be legally defensible. Alabaster & Lloyd (1980) discuss their form and applications. These authors propose that such tests take the following format:

'When the effluent is diluted X times with the specified dilution water and tested by the required procedure, no more than Y fish shall die within the prescribed period of the test'.

An example of this type of test is that prescribed by the UK Ministry of Housing and Local Government (1969). It is important to realise that tests of this type are not suitable for any other purpose (Alabaster & Lloyd, 1980).

A completely different approach to a rapid screening test for lethal toxicity is the residual oxygen bioassay (Carter, 1962; Her Majesty's Stationery office, 1983a). Fish are placed in a series of concentrations of the poison under test in sealed containers, and the dissolved oxygen content of the solution is measured when the last fish in each sample dies. The asumption underlying the test is that the more toxic the poison is, the more rapidly will the fish die, and hence the greater will be the residual dissolved oxygen at the end of the test. Plotting residual oxygen concentration against poison concentration may indicate a 'threshold' concentration at which the residual oxygen level does not differ from the control value (Vigers & Maynard, 1977). Some authors (Ballard & Oliff, 1969; McLeay, 1976; Vigers & Maynard, 1977) have claimed that results of residual oxygen bioassays are closely comparable with more conventional 48-hour or 96-hour LC50 values, at least for some poisons. The advantage of this method is, of course, that it is rapid and requires only small quantities of effluent or poison solution and minimal equipment. However, the test has never been fully evaluated and is not widely used. An obvious criticism of the rationale underlying this test is that poisons have varied effects on the respiration of aquatic animals. Those which act as metabolic inhibitors, such as cyanides, will instantly depress the oxygen uptake of the exposed animals. Poisons which act in other ways often cause increased oxygen uptake, at least in the initial stages of exposure (e.g. Skidmore, 1970; Davis, 1973) as they stimulate increased physical activity associated with avoidance reactions.

For some purposes it may be desirable to employ procedures which are more complex, rather than more simple, than the conventional procedure described early in this chapter. There are several circumstances under which aquatic animals may be exposed, for very short periods, to high levels of toxic pollutants. Examples include:

(a) Contamination of water by pesticides during and after aerial spraying.
(b) Direct application to water of pesticides or herbicides, for example for disease or weed control.
(c) Accidental, negligent or illegal discharges.
(d) Treatment plant failures.
(e) Entrainment of animals into effluent plumes or mixing zones round outfalls.

Such circumstances frequently give rise to mortalities of fish and invertebrates which regulatory authorities are required to investigate. In the design of outfalls and their associated mixing zones, predictions may need to be made of the effects on the receiving water fauna of entry or entrainment into effluent plumes. Direct application of herbicides or pesticides to water must be done in such a way as to minimise their impact on non-target organisms. For all these reasons it is important to measure the toxic effect of short exposures to relatively high levels of pollutants.

It might be assumed that the required information can easily be obtained from the conventional toxicity curve (Figs 4.2, 4.4, 4.6). For a given concentration of poison, the time required to cause half the animals to die can be directly read from the curve. However, it has been shown (Abel, 1980a, b) that this approach may seriously underestimate the lethal impact of short exposures to lethal concentrations. This is because the curve relates poison concentration and survival time for animals *exposed continuously to the poison until they die*. It thus ignores the possibility that death of the animal may be determined within a period shorter than its actual survival time. The significance of this is that the conventional toxicity curve may seriously over-estimate the duration of exposure required to cause death of the animals (Abel, 1980a, b).

The reasons for this have to do with the design of the toxicity test itself. Death of the animal will be determined when the animal has accumulated, or when its tissues have reacted with, a specific quantity of poison, the *lethal dose*. In toxicity experiments with aquatic animals, the lethal dose is rarely determined. The quantity of poison taken up by the animal will be a function of the poison concentration (hence the diffusion gradient) and the duration of exposure. The conventional toxicity test indicates only the relationship between survival time and poison concentration, and almost always shows that as concentration increases, survival time is decreased. Thus each group of animals is exposed not only to a different poison concentration but also for a different time period, so no information is yielded on the interaction of poison concentration and exposure time which determines the response of the animals.

Study of the relationship between mortality, poison concentration and exposure time requires a more elaborate experimental protocol (Abel, 1980a). Animals are exposed to each of a range of poison concentrations for each of a series of fixed exposure times, and then transferred to clean water. Subsequent mortalities are recorded at each of a series of observation times. For each observation time and each concentration of poison, log-probit graphs of percentage mortality (probits) against exposure time (log) are constructed. The exposure time causing 50% mortality within the specified observation time is read off and its confidence limits estimated by the procedure of Litchfield & Wilcoxon (1949). This process is exactly analogous to the determination of LC50 values, but the result is an estimate of median lethal exposure time rather than of median lethal concentration of poison. Values for median lethal exposure time may then be plotted against observation time (equivalent to survival time), giving for each concentration a curve which is analogous to the conventional plot of median lethal concentration against survival time, Fig. 4.11. This diagram summarises the relationship between survival time, exposure time and poison concentration, in the example given for *Gammarus pulex* exposed to the pesticide lindane.

These data indicate two significant points. First, that following a relatively short exposure to lethal levels of poison animals will continue to die over a period of at least two to three weeks, even if all survive the initial exposure. The curves shown in Fig. 4.11 do not become asymptotic to the time axis, indicating that a threshold median lethal exposure time has not been established and that it may be measurable in minutes rather than hours or days. Secondly, exposure times required for a given concentration of poison to cause mortality are considerably shorter than might be expected from the results of conventional toxicity tests. The toxicity curve for *G. pulex* exposed to lindane, derived by conventional procedures, is shown in Fig. 4.12. This indicates that when animals are continuously exposed to lindane at a concentration of 1 mg l^{-1}, half will die in about two hours, while at a lindane concentration of 0.1 mg l^{-1}, half will die in about 20 hours. However, Fig. 4.11 indicates that in fact exposure to 1 mg l^{-1} for only 15 minutes, and to 0.1 mg l^{-1} for only three hours, will eventually cause half the animals to die. It is possible by the methods described above (Abel, 1980a, b) to construct a curve showing the effect of poison concentration on say, the 20-day median lethal exposure time (i.e. the duration of exposure required to cause the death of half the animals within 20 days following the initial exposure). Similar graphs can, of course, be constructed for any specified period within the duration of the experiment. Fig. 4.13 shows such a curve, from which combinations of poison concentration and lethal exposure times can be directly read. The conventional toxicity curve is reproduced, to the same scale, on the same diagram

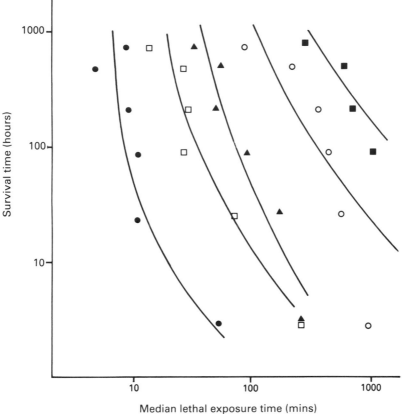

Fig. 4.11 — Relationship between median lethal exposure time and survival time for *Gammarus pulex* exposed to each of five concentrations of the pesticide Lindane. Concentrations are (left to right) 2, 1, 0.5, 0.2 and 0.1 mg l^{-1}. Confidence limits have been omitted for clarity (Abel, 1980a).

and allows easy comparison of the results of the two test procedures. Clearly the exposure time required to cause mortality is far less than the conventional procedure would indicate. Similar results have been obtained for copper (Abel, 1980b) zinc (Abel & Green, 1981) cadmium, Permethrin and cyanide (Abel & Garner, 1986) and some data are shown in Table 4.1. This more elaborate toxicity test procedure therefore provides additional information of potential use in evaluating the impact of short-term, high level exposure to pollutants, such as may arise in the kinds of circumstances outlined above.

4.1.6   Evaluating test methods and their results
The examples discussed illustrate the variety of methods available for measuring lethal toxicity. It has already been emphasised that a prime consideration governing

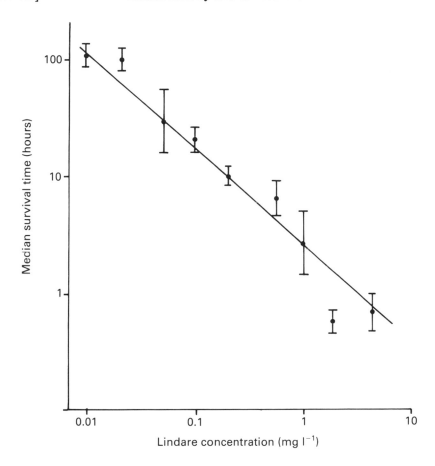

Fig. 4.12 — Toxicity curve for *Gammarus pulex* exposed to Lindane, derived by conventional procedures (Abel, 1980a).

the choice of method should be the purpose for which the results are required. Conversely, the interpretation and applications of toxicological data are circumscribed by the methods used to obtain them. A large part of the toxicological literature consists of values for 96 hour (or shorter) median lethal concentrations of poison for a wide variety of species. We have already seen that such data do not provide an adequate basis for the study of comparative toxicity, and even less so for the formulation of water quality standards. Unfortunately, in practice the biologist frequently has to make judgements, or predictions, on the basis of inadequate information. In practice, informed judgement based on cautious interpretation of limited data is preferable to no judgement at all. Therefore it is important to understand the principles underlying toxicity test procedures, in order to be able to assess the relevance, value and limitations of their results.

Finally, we may consider the question of standardisation of toxicity test methods. Variations between methods occur in the choice of test species, the environmental

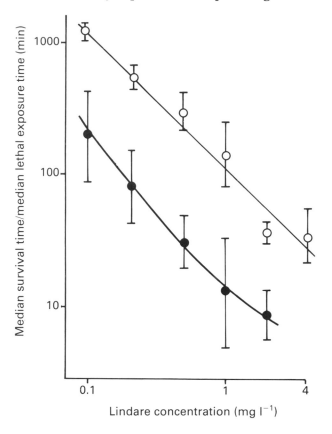

Fig. 4.13 — Toxicity curves showing median survival times (upper curve) and median lethal exposure times (lower curve) for *Gammarus pulex* exposed to a range of concentrations of Lindane (Abel, 1980a).

conditions under which the poison is tested, the treatment of the test animals before and during the test, and methods of data collection and analysis, and so on. Alabaster & Lloyd (1980) describe several standard test techniques adopted in various countries. Standardisation of procedures has obvious importance and advantages for some purposes, and has been welcomed by some authors (e.g. Sprague, 1970). It contributes to the rapid accumulation of knowledge by facilitating comparisons of results, for example between different laboratories. Since toxicity measurements may have some bearing on the legislative aspects of pollution control, some standardisation of techniques is essential. However, there are both scientific and practical objections (Brown, 1973) to excessive standardisation. No standardised procedure can be suitable for all circumstances. The variety of environmental conditions under which organisms are exposed to pollutants, and the range of species potentially at risk, are very wide. Thus there is a danger that as standard test species, or standard genetic strains of species, or standard test conditions become more widely adopted, the information obtained will become less applicable to any specific set of circumstances.

Table 4.1 — Comparison of median survival times and median lethal exposure times for *Gammarus pulex* exposed to Lindane, copper and cyanide

Poison	Concentration mg l^{-1}	Median lethal exposure time (hours)	Median survival times (hours)
Lindane	0.1	3.2	20
	0.2	1. 3	9.4
	0.5	0.5	5.4
	1.0	0.22	2.4
	2.0	0.15	0.57
Copper	0.2	1.1	9
	0.4	0.7	5.2
	0.6	0.4	4.0
	0.8	0.3	3.4
	1.0	0.2	2.8
Cyanide	3	5.0	15
	7.5	1.1	9
	15	0.75	6.2
	30	0.75	5.7
	75	0.3	2.9

[a]Lindane: 20-day median lethal exposure times (Abel, 1980a).
Copper: 15-day median lethal exposure times (Abel, 1980b).
Cyanide: 14-day median lethal exposure times (Abel & Garner, 1986).

A useful approach is to compare the performance of different test procedures to evaluate the nature and extent of variations in the results they produce. Reiff *et al.* (1979), for example, summarised the results of an interlaboratory exercise in which the toxicity of eleven detergents was tested by seven different procedures in thirteen laboratories in Germany and Britain. Despite differences in the species used and in the environmental conditions of each test method, the results were in general closely similar in the magnitude and ranking of the LC50 values. The results did, however, indicate some consistent interspecific variations in susceptibility. Studies of this kind seem to suggest that standardisation of test species and environmental conditions is less important than the application of sound methodology. As Brown (1973) concluded, 'The only standards to be applied in toxicity testing are those of good experimental techniques and sound scientific practice.'

4.2   FACTORS INFLUENCING TOXICITY

There is a great deal of published information on the comparative toxicity of pollutants to different species under different conditions, and the present discussion is a selective review of some of the more salient features of the literature. Only influences on lethal toxicity are considered here: factors affecting sublethal toxicity have been little studied. Regrettably, much of the available literature is of limited value, or at least requires cautious interpretation, largely because of the kinds of methodological limitations discussed earlier. For example, we have seen (Fig. 4.4) that comparing two LC50 values for a single time period such as 48 or 96 hours can lead to erroneous conclusions. Any biotic or abiotic influence on toxicity may affect the speed of response of the organisms as well as, or instead of, the actual quantity of poison required to kill the animals. Therefore comparisons of toxicity based on measurements of survival times, mortality rates, or fixed-time LC50 values are not adequate firmly to establish the existence or magnitude of an effect on toxicity. The importance of determining lethal threshold values is rarely more evident than when discussing comparative toxicity.

Unfortunately, much of the literature on comparative toxicity takes precisely the form of tabulations of LC50 values, or comparisons of survival times or mortality rates. Therefore the existence and magnitude of many potential influences on toxicity is not well established, despite in many cases the existence of a substantial literature.

4.2.1 The toxic properties of the pollutant

The toxicity of a pollutant under any given circumstances must be a function of its chemical structure or configuration, and quite small alterations in the poison molecule can produce major variations in toxicity. Synthetic anionic detergents are a group of compounds which illustrate this well. Alkylbenzene sulphonate detergents were widely used until the mid-1960s but, being refractory to biological waste treatment processes, were generally replaced with linear alkylate sulphonates, which only differ in having an unbranched hydrocarbon chain. The newer LAS detergents are generally more toxic than the ABS type which they replaced (Abel 1974). However, because they break down more rapidly in treatment plants they produce fewer foaming and toxicity problems in receiving waters. The degradation products of detergents are markedly less toxic to aquatic organisms than the original molecule (Swisher *et al*. 1964; Kimerle & Swisher, 1977). Numerous studies have shown that the length of the hydrocarbon chain of an anionic detergent molecule exerts a large influence on its toxicity (Hirsch, 1963; Lindahl & Cabridenc, 1978; Maki & Bishop, 1979). These studies are in general agreement that the toxicity of anionic detergents to fish, invertebrates and algae (expressed in terms of 48 h LC50 values) increase by up to one order of magnitude for each increase of 2 alkyl carbons in the detergent molecule, although very long hydrocarbon chains (C16 and above) tend to show reduced toxicity.

The toxicity of both organic and inorganic poisons is greatly influenced by the physicochemical state in which the poison is present. Heavy metal ions, for example, may exist in any of several different oxidation states; in dissolved, colloidal or particulate form; and as simple ions or as inorganic or organo-metal complexes. These various forms may have very different toxic properties. This fact explains, at least in part, the effect of some environmental conditions on toxicity.

The toxicity of copper provides a good example. Pagenkopf *et al*. (1974) recorded from published literature 96 h LC50 values for *Pimephales promelas* ranging over two to three orders of magnitude. From the chemical data provided on the environmental conditions of these experiments, they calculated the equilibrium concentrations of five possible copper species ($CuCO_3$, $CU(CO_3)^{2-}$, $CU_2(OH)_2^{2-}$, $CuOH^+$ and Cu^{2+}). The concentration of Cu^{2+} required to kill half the fish within 96 h was nearly constant, and it is apparently the major toxic species $CuOH^+$ was found to be rather less toxic, and the other copper species apparently contributed little to the toxic action of copper. In each case examined, a substantial proportion of the total copper present was complexed with carbonate and hydroxide. It is also known that complexation of copper with dissolved organic material causes a marked reduction in its toxicity. Sewage effluent, glycine, humic substances, and suspended organic matter (Brown *et al.*, 1974) and organic chelating agents such as nitrilotriacetic acid (Shaw & Brown, 1974) markedly reduce the toxic action of copper.

The influence on toxicity of the physico-chemical state, and molecular structure or configuration, of pollutants may thus have extensive implications for the conduct of toxicity tests, and for the interpretation of test results particularly when such results are to be extrapolated in the formulation of water quality standards. Lee (1973) in his review of the topic pointed out that with the increasing use of chronic, sublethal toxicity testing, concentrations of pollutants which are found deleterious in experimental conditions are sometimes equal to or less than the apparent 'natural' concentrations found in waters which sustain a healthy biota. An obvious possible explanation is that under laboratory conditions the poison may largely be present in a form different from that which predominates in natural waters. While it has long been standard practice to monitor, control and record the physical environment of toxicity tests, until recently little attention has been paid to the chemistry of the poison under test conditions. Lee (1973) recommends procedures for minimising the problems which may be associated with the influence of the chemical environment on the results of toxicity tests, in particular that in the design and execution of toxicity tests, consideration should be given to the chemistry of the test substance, and to the determination of the precise form of the substance which is responsible for the observed toxic effect.

4.2.2 The effects of environmental conditions

Since the environmental conditions may affect both the poison and the organism under test, it is not surprising that their influence on toxicity can be large. An exhaustive treatment of the enormous literature on the influence of environmental conditions is beyond the the scope of the present discussion. Rather, a selective approach has been adopted to illustrate the nature and magnitude of environmental influences on toxicity.

Water hardness is among the most important environmental influences on toxicity, and its effects are particularly well known in relation to heavy metal toxicity. Lloyd (1960) found that the concentration of zinc which was lethal to rainbow trout (*S. gairdneri*) within two and a half days varied by a factor of eight over the hardness range 12–320 mg l^{-1} as $CaCO_3$. Other metals are similarly influenced, including copper (Howarth & Sprague, 1978), cadmium (Calamari *et al.*, 1980) and silver (Davies *et al.*, 1978). Early explanations of the effect of water hardness on metal toxicity, briefly reviewed by Skidmore (1964), centred on the effects of alkaline earth ions on the physiology of the test animals. More recently, as described above, it has become clear that water hardness and other environmental variables influence the distribution of the total available metal ions between each of several inorganic complexes, which are not all equally toxic. However, that the effect of hardness on toxicity has at least partly a biological basis can be shown by acclimating fish to hard water and exposing them to the metal in soft water. Such fish are more resistant to zinc (Lloyd, 1965) and cadmium (Calamari *et al.*, 1980) than similar fish acclimated and exposed in soft water. Hardness has also been reported to affect the toxicity of poisons other than metals, for example fluoride (Herbert & Shurben, 1964). Sprague (1970) summarised the literature to date and little has since been added.

The effects of *temperature* on toxicity may be expected to be complex. Temperature influences the rate of metabolic processes, including the uptake, metabolism and excretion of poisons. Increased temperature will increase the oxygen requirements of aquatic organisms, while decreasing the solubility of oxygen in water. The properties of the poison itself may, of course, be directly influenced by temperature; for example, through its effect on the equilibrium between molecular and ionised forms. Temperature is itself an important limiting factor to aquatic organisms. Further, as Sprague (1970) points out, it appears to be particularly important in studying temperature effects to determine lethal threshold concentrations, rather than LC50 values at arbitrarily-selected observation times. This is because temperature may influence the rate of reaction of the organism to the poison, but not the actual lethal threshold concentration. For these reasons, the large literature relating to temperature effects on toxicity affords little opportunity for reliable generalisation. It appears to be generally true that at higher temperatures, the time taken for organisms to react to a given concentration of poison is reduced. Cairns *et al.*, (1975) cite numerous examples. It follows that results of short-term tests where lethal thresholds are not established, and in particular comparisons based on 48 h or 96 h LC50 values (i.e. the vast majority of reported results), are likely to be misleading and to exaggerate the magnitude of apparent temperature effects on toxicity. Reliable reports of temperature effects on lethal threshold concentrations appear to be very scarce.

These points are well illustrated by the following examples. Hokanson & Smith (1971) showed that the lethal threshold concentrations of an anionic detergent to *Lepomis macrochirus* were similar at 15°C and 25°C. However, at the higher temperature median survival times were reduced and the lethal threshold time was 20–24 h at 25°C compared with 48–51 h at 15°C (Fig. 4.14). Adelman & Smith (1972) studied the effects of temperature on the toxicity of hydrogen sulphide to goldfish, *Carassius auratus*, in tests lasting 11 days. Their results showed that lethal threshold concentrations ranged from approximately 90 μg l^{-1} at 14°C to approximately 60 μg l^{-1} at 26°C, that is hydrogen sulphide was more toxic at the higher temperature. However, the complete toxicity curves (Fig. 4.15) show clearly that the apparent increase in toxicity with temperature was much more pronounced in tests of shorter duration: corresponding 96 h LC50 values were approximately 140 μg l^{-1} and 65 μg l^{-1} respectively. An example of decreased toxicity with increasing temperature is given by Brown *et al.* (1967). The 48 h LC50 of phenol to *Salmo gairdneri* was 5 mg l^{-1} at 6°C and increased to 9 mg l^{-1} at 18°C. Toxicity curves published by these authors indicate that 48 h LC50 values were in this case close to the lethal threshold concentrations. Interestingly, at all but the lowest concentrations tested the effect of

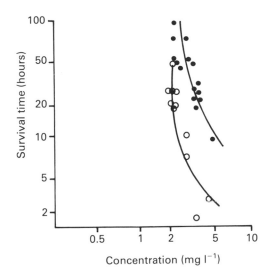

Fig. 4.14 — Effect of temperature on the toxicity of linear alkylate sulphonate detergent to *Lepomis macrochirus*. Open circles represent tests at 25°C, closed circles tests at 15°C (Hokanson & Smith, 1971).

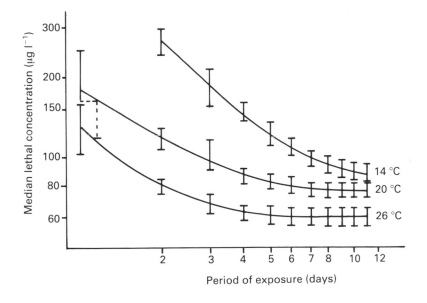

Fig. 4.15 — Effects of temperature on the toxicity of hydrogen sulphide to *Carassius auratus* (Adelman & Smith, 1972).

increased temperature was to *reduce* median survival times, that is the toxicity curves intersect in the manner illustrated in Fig. 4.7. Thus even when the effect of increased temperature is to reduce toxicity in terms of lethal threshold concentration, comparison of short-term LC50 values would lead to exactly the opposite conclusion. These examples all illustrate the danger of arbitrarily-terminated toxicity tests, and explain why, unfortunately, most of the literature relating to temperature effects on toxicity must be regarded as being of very limited value.

The effects of *dissolved oxygen* on toxicity have been less widely investigated, but in general low dissolved oxygen (DO) concentrations appear to cause an increase in the toxicity of poisons. As with temperature, the majority of reports deal with toxicity as measured by survival times, mortality rates, or tests of arbitrarily-fixed short duration, and accordingly need cautious interpretation, but there are some clear examples of the effects of oxygen concentration on threshold toxicity values. Hokanson & Smith (1971) found that the lethal threshold concentration of an LAS detergent to *Lepomis macrochirus* increased from 0.4 mg l^{-1} at 2 mg DO l^{-1} to 2.2 mg l^{-1} at 7.5 mg DO l^{-1}. Thurston *et al.* (1981) reported lethal threshold concentrations of ammonia to *S. gairdneri* ranging from 0.32 mg $NH_3 l^{-1}$ at 2.6 mg DO l^{-1} to 0.81 mg $NH_3 l^{-1}$ at 8.6 mg DO l^{-1}. Acclimation of fish to low dissolved oxygen concentrations prior to testing may influence the results of experiments. Lloyd (1960) found a small increase in the toxicity of zinc to *S. gairdneri* at low DO levels which was abolished if fish were acclimated to low DO before testing. Similar results were obtained by Adelman & Smith (1972) for *Carassius auratus* exposed to hydrogen sulphide.

Although extreme pH values are deleterious to aquatic organisms, the pH range 6–9 is generally considered acceptable to most species. Within this range, however, the toxicity of many poisons is influenced by pH. Particularly strongly affected are those poisons which dissociate into ionised and unionised fractions, of which one is markedly more toxic than the other. The best-known example is ammonia (Alabaster & Lloyd, 1980) which is more toxic at high pH values. The reason is that unionised ammonia has high toxicity and the ammonium ion has very low toxicity, and the proportion of unionised ammonia in solution increases rapidly with pH; thus the toxicity of ammonia is about ten times higher at pH 8 than at pH 7. Alabaster & Lloyd (1980) also cite cyanide, nickelo-cyanide complex, sodium sulphide and zinc as examples of poisons whose toxicity is substantially influenced by pH. It is reasonable to assume that pH is potentially an important determinant of toxicity for any poison which ionises in solution.

The effect of *salinity* on toxicity has received some attention since many pollutants are discharged to saline waters. Some poisons appear to be least toxic to fish at salinities corresponding to approximately 30–40% sea water, when the water is roughly isotonic with fish body fluids. Examples include zinc (Herbert & Wakeford, 1964), ammonia (Herbert & Shurben, 1965) and alkylbenzene sulphonate detergent (Eisler, 1965). The effect of salinity on zinc toxicity appears particularly large (Fig. 4.16) although in this case the toxicity is expressed as 48 h LC50 and not lethal threshold concentrations. Phenol toxicity to *Salmo gairdneri* increased steadily with salinity (Brown *et al.*, 1967), 48 h LC50 values ranging from 9 mg l^{-1} in fresh water down to 5 mg l^{-1} in 60% sea water. Cadmium is reported to increase in toxicity to the estuarine fish *Fundulus heteroclitus* (Eisler, 1971) as salinity increases. There is no

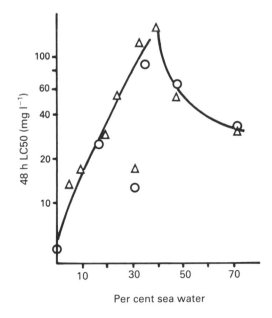

Fig. 4.16 — Effect of salinity on the toxicity of zinc to *Salmo gairdneri* (Herbert & Wakeford, 1964).

generally accepted hypothesis to explain salinity effects on toxicity. Herbert & Wakeford (1964) suggested that zinc was less toxic to fish in isotonic medium because of the reduced importance of osmoregulatory stress during intoxication. Skidmore (1970) showed that trout poisoned by zinc in fresh water maintained normal osmotic and ionic balance, and that death was due to asphyxiation associated with gill damage. This does not mean, however, that osmoregulatory requirements have no influence in zinc poisoning. It has been suggested (Abel & Skidmore, 1975) that the gill damage associated with poisoning by zinc and other pollutants is a consequence of a rearrangement of the gill epithelium to preserve an osmotic barrier in the face of a rapid loss of viable epithelial cells due to the action of the poison. Thus the gill damage, and consequent tissue hypoxia, may be less extensive for fish in isotonic medium, though this has never been investigated. Poisons which are more toxic in hypertonic medium may be so because teleosts drink copiously in hypertonic medium, and thus may accumulate poison more rapidly. The effect of salinity on chemical speciation of the toxicant (see above) may also be relevant.

4.2.3 Combinations of poisons
An important environmental variable which may influence the toxicity of a poison is the presence of other poisons. Although most toxicological investigations involve the study of a single pollutant, the biota of polluted waters are usually exposed to several pollutants simultaneously.

Sprague (1970) has pointed out that the terminology widely used to describe the behaviour of poisons acting simultaneously is confused and potentially misleading. In particular, the terms *synergism, potentiation* and *antagonism* have been defined and used in different ways by different authors, and their use is best avoided. Sprague proposed a system of nomenclature based on that of Gaddum (1948). This scheme is

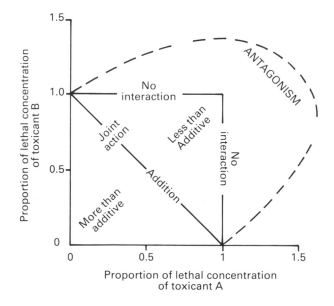

Fig. 4.17 — Diagram showing terms used to describe the combined effects of two pollutants (Sprague, 1970). For full explanation, see text.

shown in Fig. 4.17 and is the one adopted here. Sprague's description of this scheme of nomenclature is as follows.

'The diagram represents the combination of two toxicants. The axes represent concentrations. The concentration of 1.0 toxic unit of toxicant A produces the response (death in this case) in the absence of toxicant B, and 1.0 unit of B will do the same in the absence of A. If the response is produced by combinations of the two toxicants represented by points inside the square, the toxicants are helping one another; this is called *joint action* which may be further broken down in three special cases, as follows. If the response is just produced by combinations represented by points on the diagonal straight line (e.g. 0.5A + 0.5B) the effects are said to be *additive*. If the response is produced by combinations falling in the lower left triangle (e.g. 0.5A + 0.2B) the effect is *more-than-additive*. If in the upper right triangle (e.g. 0.8A + 0.7B) the toxicants are still working together in joint action but are *less-than-additive*. Those combinations falling exactly in the upper and right boundaries of the square show *no interaction* between the toxicants. For example, if 1.0 unit of A is required to just produce the response, no matter what concentration of B, below 1.0 unit, is present, then A is causing the response and B is neither helping nor hindering. If more than 1.0 unit of A is required to just produce the effect, because of the presence of B, this is *antagonism*, with B antagonising the effect of A. That combination of concentrations would fall on some point to the right of the square (e.g. 1.5A + 0.5B). Any combination of concentrations which would fall outside the square would represent antagonism, loosely represented by the broken curved line.'

A useful method for measuring the toxicity of poisons in combination has been described by Brown (1969) and Sprague (1970). The concentration of a pollutant can be expressed as toxic units, that is as a proportion of its lethal threshold concentration, or of some approximation of the lethal threshold concentration such as 48 h or 96 h LC50. Thus:

$$\text{Toxic units} = \frac{\text{Actual concentration of poison present}}{\text{Lethal threshold concentration}}.$$

Thus for any poison, one toxic unit is equal to its lethal threshold concentration. To test the toxicity of a mixture of poisons, the following procedure may be employed. Assume we wish to test the effect of two poisons acting simultaneously. Having determined the lethal threshold concentration of each poison individually, test organisms are then exposed to a mixture of the poisons which contains, say, 0.5 toxic units of each poison (i.e. half the lethal threshold concentration of each poison). If the poisons are additive, the mixture should cause half the test organisms to die. If significantly more than half die, the poisons are more-than-additive. If significantly fewer than half die, the poisons are less-than-additive. Obviously it is feasible using this technique, to investigate the effects of combinations containing more than two poisons, and of mixtures in other than equal proportions. Marking (1977) proposed a modification of this approach, whereby the toxicity of chemicals in combination may be expressed in terms of a single numerical value. Index values of zero represent additive toxicity; values significantly above, or below, zero represent more-than-additive and less-than-additive toxicity respectively. The method includes a simple test of significance for values near zero. An alternative approach (Calamari & Marchetti, 1973) is to construct complete toxicity curves for the poisons individually and in combination, converting values on the concentration axis of the graph to toxic units (Fig. 4.18).

There are several examples in the literature of both additive and more-than-additive toxicity. Mixtures of copper and zinc (Lloyd, 1961; Sprague & Ramsay, 1965); copper and phenol; copper, zinc and phenol; and copper, zinc and nickel (Marking, 1977; Brown & Dalton, 1970) have all been reported to be simply additive. More-than-additive toxicity has been reported for anionic detergents with copper or mercury (Calamari & Marchetti, 1977) the piscicide rotenone with sulphoxide or piperonyl butoxide, and the organophosphate insecticides malathion and Delnav (Marking, 1977). As in other types of toxicological study, it appears that tests of short duration where threshold toxicity values are not established are likely to

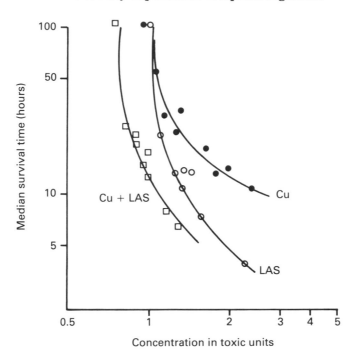

Fig. 4.18 — Toxicity of copper (Cu), detergent (LAS) and a copper-detergent mixture (Cu+LAS) to *Salmo gairdneri*. Poison concentrations are expressed in toxic units. (Calamari & Marchetti, 1973).

produce misleading results. Sprague (1970) cites examples which illustrate that while threshold toxicities may be simply additive, survival times in strong mixtures may be shorter than expected on the basis of simple additive toxicity.

Examples of less-than-additive toxicity seem to be rare. Calamari & Marchetti (1973) reported that a mixture of equal proportions (in toxic units) of copper and the non-ionic detergent nonylphenol ethoxylate showed a less-than-additive toxicity to *S. gairdneri*, but this conclusion was based on a comparison of observed and expected survival times. One circumstance in which less-than-additive toxicity sometimes occurs is when a component of a mixture is present as a small fraction of the whole mixture. Thus Brown *et al.* (1969) found that the observed toxicity of a mixture of ammonia, phenol and zinc to rainbow trout was significantly less than the predicted toxicity when zinc contributed about 0.75 toxic units and the remaining 0.25 toxic units was roughly equally distributed between the phenol and the ammonia.

In the case of ammonia, Lloyd & Orr (1969) demonstrated that concentrations below 0.12 toxic units exerted no effect on the water permeability of the fish. If toxic levels of ammonia exert their effect at least partly by increasing water permeability of the fish, as is apparently the case, this finding suggests an explanation for the non-contribution of low concentrations of ammonia to the toxicity of mixtures. For most poisons, however, no such detailed knowledge of their mode of actions exists, and their behaviour in mixtures can be predicted only on the basis of empirical findings without any explanation of the underlying mechanism. Nevertheless a series of attempts by British workers to predict the toxicity of mixtures, and to test such predictions against field observations of the fishery status of polluted rivers, have been reasonably successful. Sprague (1971) reviews some early attempts and more recent examples include those of Alabaster *et al.* (1972), Solbé (1973) and Garland & Rolley (1977). In British rivers and commonest toxic pollutants are ammonia, phenols, cyanide, zinc and copper. For those poisons, summation of the fractional toxicities expressed in toxic units as described above gives generally good agreement between predicted and observed toxicities in laboratory experiments. The application of this technique in field conditions was discussed in Chapter 1.

4.2.4 Fluctuating concentrations

The concentrations of pollutants in receiving waters are rarely constant; they fluctuate, often quite widely and with rapid periodicity. Consequently the effects of fluctuations of poison concentration are of some interest. Additionally, even in the best-designed toxicity tests some fluctuation in the poison concentration is occasionally unavoidable, and information on the response of test animals to fluctuating concentrations would also be of value in this context. Surprisingly, there have been relatively few investigations of the point. Brown *et al.* (1969) determined the 48 h LC50 of ammonia, zinc, and an ammonia-zinc mixture to *Salmo gairdneri*. They then exposed replicate groups of fish to a constant 48 h LC50 of each poison, and to fluctuating concentrations such that the mean concentration was a 48 h LC50 but the actual concentration varied between 0.5 and 1.5 times the 48 h LC50 at intervals of one to four hours. In the majority of cases there was no significant difference between the survival times of fish exposed to constant concentrations and those exposed to fluctuating concentrations, suggesting that fluctuations of ±50% of the nominal concentration do not affect the response of the fish. However, in the case of ammonia, where the periodicity of the fluctuations was two hours rather than one, the median survival time was significantly reduced. Thurston *et al.* (1981) reported similar findings, also using trout exposed to ammonia. In these experiments fish were alternately exposed to ammonia and clean water, the periodicity of the fluctuations being six or 12 hours. On the basis of 96 h LC50 values for stable and fluctuating concentrations, the fish were more sensitive to the fluctuating concentrations, than to a constant concentration equivalent to the mean value. One explanation of this (Brown *et al.* 1969) is that when the periodicity of fluctuation is long, fish will suffer irreversible damage during the exposure to the peak concentration which cannot be compensated for during exposure to the lower concentration. This argument corresponds to that advanced earlier to explain the difference between median lethal exposure times and median survival times; exposure of animals to a lethal concentration for a fixed time followed by transfer to clean water can be considered an extreme example of exposure to a fluctuating concentration of poisons.

4.2.5 Biotic factors influencing toxicity

Animals may show both interspecific and intraspecific variation in susceptibility to pollutants. It is beyond the scope of the present discussion to review in detail all the relevant literature. However, it is possible to draw attention to some of the more salient general findings regarding the way in which the biological characteristics of the test organisms influence their response to toxic pollutants.

Among fishes, interspecific variation in susceptibility is probably smaller than was once thought, and may be less important than variation due to environmental conditions. Because of the large influence of environmental conditions on toxicity, it is arguably unwise to place too much reliance on comparisons between the results of different investigators who have measured toxicity under different environmental conditions. In many studies, environmental variables which are known to have a large influence on toxicity have not been measured or specified; and there have been relatively few studies in which several species have been tested under similar conditions.

Thatcher (1966) found that 96 h LC50s of an alkylbenzene sulphonate detergent to eleven species ranged from 7.7 mg l^{-1} to 22 mg l^{-1}. For a linear alkylate sulphonate detergent tested against five species, 96 h LC50s varied from 3.3 mg l^{-1} to 6.4 mg l^{-1} (Thatcher & Santner, 1967). Lethal threshold concentrations of the fungicide captan to three fish species ranged from 29 to 64 μg l^{-1} (Hermanutz et al., 1973). The 96 h LC50 of hydrogen cyanide to five species varied from 57 to 191 μg l^{-1} (Smith et al., 1978). Interspecific variations in short-term LC50s of about threefold to fourfold appear to be typical for many types of poison (Sprague, 1970). Larger differences have occasionally been reported, particularly for pesticides. Eisler (1970) determined the 96 h LC50 of twelve insecticides for seven estuarine species. Interspecific variations for endrin were 0.05–3.1 μg l^{-1}; for dieldrin, 0.9–34 μg l^{-1}; for heptachlor, 0.8–194 μg l^{-1}; and for malathion, 27–3250 μg l^{-1}.

The importance of determining lethal thresholds, rather than relying on short-term LC50 values, is again evident from Ball's (1967a) studies on the relative susceptibility of fish species to ammonia. Although Salmonid fishes are widely considered to be more sensitive to pollutants than coarse fish, he showed that the lethal threshold concentration of ammonia for *Salmo gairdneri* was the same as that for three Cyprinid species (*Abramis brama*, *Rutilus rutilus* and *Scardinius erythrophthalmus*). However, the trout reacted far more rapidly to the ammonia than did the cyprinids, so that in tests of short duration they would appear to be more sensitive. Ball (1967c) also measured the toxicity of zinc to trout and four coarse fish species. In this case the trout were markedly more sensitive: the threshold LC50 value for trout was 4.6 mg l^{-1}, significantly lower than the corresponding values for *Rutilus rutilus*, *Abramis brama* and *Perca fluviatilis*, which lay between 14.3 and 17.3 mg l^{-1}. The gudgeon, *Gobio gobio*, had a seven-day LC50 (not a threshold value) of 8.4 mg l^{-1}. As seen earlier, the toxicity of both zinc and ammonia is greatly influenced by environmental conditions, and it is likely that variation in susceptibility between trout under different environmental conditions is actually greater than that between trout and other fish species. The results of Smith et al. (1978) indicate that environmentally-induced variability in the toxicity of cyanide to fishes is of a similar magnitude to interspecific variability.

A further point arising from Ball's (1967a) study has been referred to earlier but is also relevant here. It appears that some species may show much greater individual variation is susceptibility to a poison than others. Thus in one of Ball's experiments, the five-day LC50 of undissociated ammonia to roach (*Rutilus rutilus*) and rudd (*Scardinius erythrophthalmus*) were identical, but the slopes of the probit lines differed, indicating that roach were more variable in response than rudd. The practical implications relate to the application of toxicity data to field observations, and to the formulation of water quality standards. In an example given by the author, two-thirds of the five-day LC50 of ammonia would kill only 1% of a rudd population, but 16% of a roach population. This example again illustrates the advantages of a full analysis of toxicity test data, and the limitations of comparative studies based on simple determinations of short-term LC50 values.

Several authors have undertaken comparative studies of the susceptibility of invertebrate species to pollutants. In most of these, however, authors have chosen representative species from each of several orders, classes or phyla. Not surprisingly, the susceptibility of such 'representative' species shows some very wide variations, often of two or three orders of magnitude in terms of short-term LC50 values, in comparison with the phylogenetically more uniform fishes (e.g. Bell, 1971; Rye & King, 1976; Gaufin *et al.*, 1965). Such information is of great use to pollution biologists, and a large literature is developing which is beyond the scope of this discussion to review in detail. However, there have been some studies of the comparative toxicity of poisons to fairly closely-related species and it is useful to consider some examples in the present context. For five genera of oligochaetes exposed to four poisons, variation between 96 h LC50 values was generally less than one order of magnitude, although the two marine genera tested appeared highly resistant to the lethal action of cadmium (Chapman *et al.*, 1982). There were only small differences between the 72 h LC50 values of copper tested against four species of Daphnia (Winner & Farrell, 1976). Sanders (1970) measured the toxicity of several herbicides to six crustaceans representing four orders. In general, variations in 48 h LC50 values between species were within one order of magnitude or less, but some species displayed considerable resistance to some or all of the herbicides, and some herbicides elicited interspecific variations in 48 h LC50 values spanning three orders of magnitude. These and similar examples from the literature indicate that it is not as yet possible to draw firm conclusions about the relative susceptibility of invertebrate species to poisons, and there is a need both for further research and for detailed review of existing information.

4.2.6 Intraspecific variation
The effects of size and age of fish on their susceptibility to poisons has been the subject of surprisingly few systematic investigations, although the point is important both for the practical application of toxicological data and because laboratories which routinely carry out toxicity tests cannot always obtain animals of standard size and/or age. In most reports, fish of different sizes have also been of different ages, so the separate influences of these two variables remains largely unknown. Nevertheless it is common practice to standardise as far as possible on the age and size of animals, and to restrict the size/age range of specimens in a test as narrowly as is practicable.

Adelman *et al.* (1976) measured the toxicity of pentachlorophenol to goldfish (*Carassius auratus*) and fathead minnows (*Pimephales promelas*) of different sizes and ages. Differences in threshold LC50 values were small and probably of no practical significance. However, the range of sizes and ages tested was small, for example, from four to 14 weeks age and 13–30 mm in length for the fathead minnow.

Kumaguru & Beamish (1981) found a more marked effect of size on the toxicity of the pesticide Permethrin to *Salmo gairdneri*. The 96 h LC50 values ranged from 3 μg l^{-1} for fish weighing 1 g, to 287 μg l^{-1} for 50 g fish. In contrast Stendahl & Sprague (1972) reported that small rainbow trout (1.5 g) were more resistant than larger (12 g) fish to vanadium, although lethal threshold concentrations were not established for the smaller fish and the difference in susceptibility is probably small.

The young of most species show a clear division of their life cycle into distinct stages, and there have been several investigations of the relative susceptibility to poisons of the different life stages of fish. (Toxicity tests spanning whole life cycles are discussed later.) Skidmore (1965) found that zebrafish (*Brachydanio rerio*) were most susceptible to zinc between four and 13 days after hatching. Eggs were considerably more resistant than newly-hatched fry, and after 13 days resistance increased rapidly. This seems to be a general pattern, similar results having been reported for species exposed to detergents (Marchetti, 1965), hydrogen sulphide (Smith & Oseid, 1972) and heavy metals (Chapman, 1978).

An important potential source of intraspecific variation in susceptibility to poisons is genetic variation between different strains or populations of a species. Many waters have been subjected to pollution for a period equivalent to many generations of the organisms living there, and it may be expected that natural selection would lead to an increase in the tolerance of populations living in polluted waters. Thus natural populations may in fact be more tolerant of pollutants than those used in laboratory experiments, which are usually inbred strains or acquired from unpolluted habitats. It is important, however, to distinguish increased resistance due to genetic adaptation, from that due to acclimation effects (i.e. long-term exposure to low levels conferring increased resistance to a subsequent high-level exposure) which are not genetically determined and of which a few examples exist in the literature (Sprague, 1970). Also, since such differences in tolerance are likely to be relatively small, once again the use of short-term LC50 values or median survival times at fairly high concentrations are not really adequate measures of toxicity. Thus examples of genetically-acquired resistance to poisons are few. Brown (1976) reported that the invertebrates *Asellus meridanus* from metal-polluted streams were more resistant to copper and lead than animals from clean streams, and that the resistance persisted to the F_2 generation. Rahel (1982) reported that perch (*Perca flavescens*) from acid lakes were more tolerant to low pH than those from alkaline lakes. He did not show that the difference was genetically based, but acclimation experiments failed to produce increased tolerance in fish from alkaline lakes. Swarts *et al.* (1978) measured the resistance of several strains of brook trout (*Salvelinus fontinalis*) to low pH and investigated the effects of genetic and environmental influences on resistance. They found that fish could not be acclimated to low pH, that strain differences in resistance to low pH were detectable, but that an attempt to produce a resistant strain by breeding from fish which had survived exposure to low pH was unsuccessful.

4.3   SUBLETHAL TOXICITY

The fauna of polluted waters are more commonly exposed to relatively low concentrations of poison for long periods, rather than to levels of pollution which will cause rapid mortality. Therefore it is important to study the effects on aquatic organisms of exposure to sublethal levels of pollution over periods which represent at least a substantial proportion of their life cycle. The historical predominance of acute lethal toxicity studies has therefore often been criticised as being of limited relevance to real situations, but is due less to the failure of toxicologists to appreciate the point than to the many technical and conceptual difficulties involved in the measurement of sublethal toxicity.

Relatively few species can be satisfactorily maintained in the laboratory for long periods and fewer still can complete their life cycle under such conditions since their environmental requirements are complex, unknown, or both. Maintenance of constant experimental conditions for periods which may exceed a year is expensive of human and physical resources, and the longer an experiment continues the greater is the chance of failure due to accident or equipment malfunction. Therefore the preferred species for research of this kind are those which can be cultured in the laboratory, are indigenous to the geographical area in which the study takes place, have reasonably short generation times and which display a more or less representative response to a wide variety of poisons; that is they should not be unusually resistant or unusually sensitive to particular poisons or categories of poison. Unfortunately in many regions few, if any, species which meet all these requirements are available.

Assuming that purely technical difficulties can be overcome, the problem remains that whereas the death of an organism is an unequivocal and easily-identifiable response to toxic action, criteria of sublethal toxic effects are less easy to define and, as will be seen, if they are recognised their biological significance is frequently difficult to assess. Nevertheless the meaningful application of data from lethal toxicity studies to many practical pollution problems is difficult, if not impossible, in the absence of some information on sublethal toxicity. Therefore a wide variety of approaches to the measurement of sublethal toxicity has been employed, and some of the more important ones are reviewed here.

4.3.1 Single-species toxicity tests

Because of the practical problems of maintaining animals under experimental conditions for very long periods, early studies on sublethal toxicity generally relied on the use of histological, pathological, biochemical, haematological, physiological or behavioural criteria of toxic effect in experiments lasting weeks rather than months. Sprague (1971) has reviewed some examples of these studies. More recently, the ever-increasing range of analytical and diagnostic techniques devised by biochemists and clinical chemists have been widely applied or adapted to the detection of sublethal toxic responses of fish to pollutants. Table 4.2 lists some of the techniques which have been used. The list is a representative rather than an exhaustive one, and no attempt is made here to review this aspect of pollution toxicology in detail. Rather, the discussion will focus on some conceptual and practical difficulties raised by the very diversity of criteria which have been employed.

The objective of experiments such as those represented in Table 4.2 is essentially to determine whether animals exposed to sublethal levels of pollutant are healthy or not. It might be thought that with the aid of the wide range of modern diagnostic and

Table 4.2 — Some examples of physiological and biochemical variables which have been used as indicators of sublethal toxic effect on fish. The list is representative rather than exhaustive and further criteria drawn from earlier work are reviewed by Sprague (1971). For full bibliographic details see Abel (1988)

Variable

Cardiovascular physiology
Heart rate
Arterial PO$_2$
Ventilation rate
Cough frequency
Opercular and buccal pressure
Ventilation volume
Oxygen consumption
Oxygen utilisation
Ventilation frequency

Haematology
Haematocrit
Haemoglobin content
Methaemoglobin content
Total blood cell counts
Erythrocyte counts
Differential leucocyte counts
Erythrocyte ATP concentration

Blood metabolite levels
Lactate
Glucose
Sodium
Chloride
Cortisol
Serum proteins
Pyruvate
Osmolality
Cholesterol
Other electrolytes

Enzyme assays (various tissues)
Lactate dehydrogenase
Glutamicoxaloacetic transaminase
γ-Aminolevulinate dehydrase

Others
Urine pyruvic acid content
O$_2$ consumption of tissue homogenate
Electrophoretic patterns of serum proteins
Body moisture content
Body lipid content
Body protein content
Liver glycogen content
Liver phenylalanine content
Urine excretion rate
Swimming performance
Food conversion efficiency
Locomotor activity

---

analytical techniques, the state of health of the test animals can be relatively easily determined, but this is not necessarily the case. For example, in order to state that a particular value is abnormal it is necessary to know the normal range for that particular variable, and the way it is affected by the physiological status and environmental history of the animal. For aquatic animals, including fish, such detailed knowledge of their biochemistry and physiology is generally lacking. Thus although it is possible to say that a particular value is statistically different from that of the control animals, it cannot readily be inferred that the change has any ecological consequences. The 'abnormal' value may represent not damage to the fish, but a metabolic adjustment well within the animal's ability to compensate for varying environmental conditions, which are a normal feature of aquatic life and to which many aquatic animals have a wide range of tolerance. Mount & Stephan (1967a) succinctly stated the difficulty thus: 'An exposure causing death is obviously significant, but even the best fish physiologist would have difficulty establishing that a 10 per cent reduction in haematocrit would result in an undesirable effect on a population'.

A more detailed exposition of the problem was given by Lloyd (1972) with the aid of the diagram reproduced in Fig. 4.19. The diagram shows the hypothetical

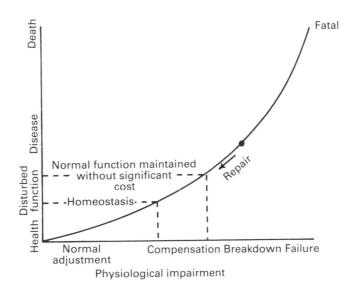

Fig. 4.19 — A possible relation between physiological impairment following increasing exposure to pollutants and the consequent disability of the fish (Lloyd, 1972).

relationship between physiological impairment following exposure to pollutants, and the consequent disability of the fish. Measured values of physiological or biochemical variables, or alterations in the behaviour of the animal or in the histological appearance of a tissue, may represent conditions within the areas of the graph marked 'homeostasis' or 'normal function maintained without significant cost', even though they may be statistically different from control values. The toxicologist's problem is to distinguish the point at which the value of a measured variable deviates so far from the control that it falls outside these zones. Unless this is clearly established, any change in the value of a measured variable is not necessarily an indicator of sublethal toxic effect. An example is provided by the report of Grant & Mehrle (1973) on the effect of exposure to sublethal levels of endrin on 19 physiological and biochemical variables in the rainbow trout. Although statistically significant differences occurred in 12 of these, the authors showed that nine out of 16 blood serum variables showed similar changes when the fish were subjected to moderate exercise, thus casting doubt on their usefulness as indicators of toxic effect.

Thus the validity of the approach to sublethal toxicity which is implicit in much of the work published in the last 20 years or so is questionable. Essentially this implicit approach has been to measure as many variables as possible, and to seek to determine the 'No observed effect concentration' (NOEC), that is the highest concentration which has no observable effect on any of the variables measured. The 'maximum acceptable toxicant concentration' (MATC) is thus determined as lying between the NOEC and the next highest concentration tested. This rationale may be criticised on several grounds. Firstly, as we have seen, a statistically significant difference in a measured variable between exposed and control fish does not imply that sublethal toxicity has occurred, unless it can be shown or at least reasonably expected that the change has actual or potential ecological significance. Secondly, there is a certain arbitrariness under this protocol in the decision as to whether or not a particular concentration exerts a sublethal toxic effect. If, for example, in an experiment 20 variables are measured, it follows that another 20, or 50, or 100, have not been measured. Any of these might, if they had been measured, have shown a difference from the control value. Thus the NOEC is determined partly by the choice of variables to be measured during the experiment. Further, there is no general agreement on what variables should be measured, so comparisons of results from different sources are difficult. Practising scientists will recognise that some variables are measured because they are easy to measure, some are selected in order to follow precedent, and some because of the availability of equipment or skilled personnel capable of making the measurement. Of course some measurements are made because there is a sound biological reason for making them, and as will be seen later it is becoming possible at least for some pollutants to identify on a rational basis specific and useful indicators of sublethal toxic effect. Finally, the use of the NOEC as the end-point of the experiment is a statistical absurdity since, as Skalski (1981) pointed out, it depends upon the non-falsification of the null hypothesis, a procedure which cannot be carried out with confidence (in the statistical sense). Such criticisms are not to deny the usefulness of the existing literature and practices in the study of sublethal toxicity. Rather, they are a reflection of the fact that the methodology is in a relatively early stage of development.

It is generally accepted that a pollutant effect on growth, reproduction or development of a species is an unequivocal criterion of sublethal toxic effect, since its ecological significance is reasonably clear. The first successful toxicity tests over a complete life cycle of a fish species appear to be those of a group of American workers (e.g. Mount & Stephan, 1967a, b) using the fathead minnow, *Pimephales promelas*. Since then, tests have been successfully carried out with about half a dozen fish species, mainly North American or small tropical species. It remains true that the number of species with which it is practicable to carry out such tests is, at the present time, a very small proportion indeed of the aquatic fauna as a whole. Apart from reproduction, it is clearly easier to carry out investigations on the effects of pollutants on growth rates, using a wider range of species. However, there are examples in the literature which show that growth rate is not necessarily a very sensitive indicator of sublethal toxic effect (Sprague, 1971), and even some examples of growth apparently being stimulated by sublethal concentrations of poison (e.g. McLeay & Brown, 1974).

Obviously experiments conducted over the whole, or a substantial proportion, of the life cycle of a species are both expensive and time-consuming, and it is not feasible to test all poisons, species and environmental conditions using such procedures. Consequently it remains important to develop and evaluate rapid methods for measuring sublethal toxicity. One approach is the so-called 'critical life stage bioassay'. Analysis of the results of a large number of partial- and complete-life-cycle tests with various species and poisons shows that in the majority of cases, the early embryo and larval stages are the most sensitive part of the life cycle, and an estimate of the MATC based solely on the response of the embryo-larval stages generally lies very close to the value obtained when the whole life cycle is considered (Macek & Sleight, 1977; McKim, 1977). Thus the duration and scale of experiments can be considerably reduced, and the critical life stage bioassay (sometimes called the 'embryo-larval test') is becoming an important technique. It also offers the possibility of increasing the range of species available for testing, since there are several species (e.g. many salmonids) whose eggs and early life stages can be maintained in the laboratory but which are difficult or expensive to maintain throughout an entire life cycle. Nevertheless there remain many species of interest which cannot be used, or which are only available during a relatively short period of the year. Thus interest remains strong in alternative criteria of sublethal toxicity.

For the reasons outlined above, such criteria should preferably be specific responses to the pollutant, that is related to the poisons's mechanism of toxic action, rather than non-specific responses which may merely represent physiological adjustment to new, but perfectly tolerable, environmental conditions. As we know relatively little about mechanisms of toxic action in fish and aquatic invertebrates, examples of such criteria are relatively rare. The work of Lloyd & Orr (1969) is a well-known case. Their investigations showed that sublethal concentrations of ammonia caused an increased urine flow rate in rainbow trout, apparently due to an increase in the water permeability of the fish. The increase of urine production was proportional to the ammonia concentration, and lethal toxicity began to occur when the urine flow rate reached the maximum of which the fish were capable. Thus the authors could predict that any factor which affected water balance was likely to affect ammonia toxicity. For example, fish which are handled immediately before testing show increased resistance to ammonia, since the diuresis caused by the handling confers an advantage on such fish in coping with the water influx caused by ammonia.

Another promising example is the finding that vertebrates, including fish, and many invertebrates produce a group of low-molecular weight proteins, metallothioneins, in their tissues when exposed to sublethal concentrations of heavy metals

(Kagi & Nordberg, 1979) The induction of metallothioneins therefore indicates that animals have been exposed to abnormally high levels of heavy metals, and the measurement of metallothionein levels in tissues has been used to assess the extent of metal pollution to which field populations are subjected (Roch *et al.*, 1982). However, the link between elevated metallothionein levels and ecologically-significant parameters such as growth rate or reproductive success has not yet been demonstrated.

Mehrle & Mayer (1980) in a brief review of clinical tests in aquatic toxicology, drew attention to a promising series of investigations involving study of the effects of poisons on biochemical processes related specifically to growth in fish. They argued that growth in fish is the culmination of a series of biochemical processes which should show changes *before* any effect on growth rate is detectable by conventional measurements of weight and length. They showed that several organic toxicants affected the vertebral collagen content of fish, and the proline and hydroxyproline concentrations in the collagen. These biochemical variables were correlated with fish growth rate, and were more sensitive indicators of toxic effect than measurement of growth rate itself.

A group of enzymes known as mixed-function oxidases (MFO) are widely found in vertebrates, including fishes, and have been reported in some invertebrates. They show some potential as indicators of sublethal toxic effect, and possibly as a means of monitoring pollutant effects in field populations (Addison, 1984). The normal function of MFO appears to be associated with the metabolism of steroid hormones, but they have been found to be produced at elevated levels in animals exposed to some pollutants, particularly aromatic hydrocarbons and halogenated hydrocarbons. Metabolites of these important pollutants appear to resemble those of some natural steroids, and to induce MFO synthesis. Thus elevated MFO levels may be used to indicate exposure of the animal to pollutants. The technique and its applications are not, as yet, well developed; for example, the link between MFO synthesis and actual toxic effect is not generally established. Mixed-function oxidase production may, for example, represent an adjustment whereby 'normal function is maintained without significant cost' (see Fig. 4.19). Another promising but as yet poorly-developed line of enquiry is the application of genotoxicological techniques to the study of water pollution. This approach is based on the finding that in virtually all organisms, exposure to poisons causes alterations in the DNA of the cells, for example an increase in the frequency of occurrence of DNA strand breaks. Such damage to DNA occurs during the normal functioning of the organism, and a mechanism for repair of damaged DNA exists in healthy cells. However, exposure to poisons can increase the extent of DNA damage beyond the capacity of the repair system, at which point toxic effects may be manifested. Genotoxic effects appear to be relatively unspecific; indeed many substances which are not normally considered poisons (e.g. some constituents of natural, uncontaminated foodstuffs) have been found to induce detectable DNA damage. This suggests that these genotoxicological techniques are remarkably sensitive, and this sensitivity is potentially useful, for example in monitoring pollutant effects in field populations. The extraction of DNA from tissues and the study of its properties can nowadays be performed routinely in well-equipped biological laboratories, and some attempts have been made to apply this technique to the study of pollutant effects in aquatic organisms (e.g. Zahn *et al.*,

1983). Again, it is not yet clear whether a given level of DNA damage can be associated with more conventional measurements of sublethal toxic effect, though arguably it is reasonable to assume that any increase in the rate or extent of DNA damage in a population is deleterious. Undoubtedly, biochemical methods of assessing pollutant effects will be further developed in the near future.

4.3.2 Experimental ecosystems

An alternative approach to the study of sublethal toxicity is that of the experimental ecosystem. Instead of exposing a population of a single species to the pollutant, populations of two or more species, frequently representing different trophic levels, are maintained and exposed together to the pollutant. In part this approach represents an attempt to make the experimental situation more realistic, and hence more directly applicable to the field situation. For example, in the field the ecology of a species is governed not only by its relationship with the physical and chemical environment and by purely endogenous population processes, but also by its relationship with other species (prey, predators, competitors, parasites) with which it shares its habitat. Thus the ecological effect of a pollutant on a species does not depend only upon the properties of the species itself; its influence may be accentuated or attenuated depending upon the nature of interspecific relationships which exist in the environment. Experimental ecosystems are both useful and necessary for studying these and other phenomena, including the distribution of pollutants between water, sediments and living tissues; the biodegradation of pollutants; and the accumulation of pollutants by living organisms from their environment, and the passage of pollutants through the food chain.

Experimental ecosystems vary in their scale and complexity. Generally the smaller-scale systems offer the advantages of more precisely-controlled experimental conditions, but are at best gross simplifications of real systems. Larger-scale systems more closely represent natural systems, but clearly cannot be so precisely controlled and experiments may be difficult or impossible to replicate. The smallest systems are more-or-less enclosed vessels containing populations of micro-organisms or plankton. On a larger scale, experimental aquaria or ponds ranging in size from a few litres to a few cubic metres have been used (Giddings, 1983). Experimental river channels, both on a laboratory scale and outdoors, have been widely used (e.g. Arthur *et al.*, 1982; Watton & Hawkes, 1984). Finally, there have been attempts to isolate large water masses in lakes and coastal marine areas, to study the processes which occur within, as it were, a representative sample of the natural environment (Steele, 1979; Davies & Gamble, 1979).

Since they vary widely in their scale, complexity and objectives perhaps the only generalisation which can be made about experimental ecosystems is that they can provide valuable information which bridges the gap between laboratory and field studies. Large-scale systems are, however, expensive to maintain, and their use is likely to be confined to aiding the interpretation of the results of more economical methods of study. (See also section 7.2.3).

4.3.3 Bioaccumulation

Bioaccumulation is an aspect of sublethal toxicity which has received much attention. Pollutants may, over long time periods, accumulate in tissues to levels which may be harmful to the organism. Since many aquatic species are utilised for human consumption, the public health significance of toxic substances accumulated in their tissues is obvious. Many national and international agencies set concentration limits for pollutants, particularly heavy metals, in tissues for human consumption, and promote research and monitoring programmes. Study of the uptake, metabolism and excretion of pollutants, and of their distribution in the various body organs and tissues, makes an important contribution to understanding their mechanisms of action. Levels of pollutants in the tissues of living organisms are widely used to indicate the degree of contamination of the waters in which they live, particularly when the pollutants are present only intermittently or in very low concentrations, making chemical analysis of the water difficult. Finally, many poisons, particularly heavy metals and refractory organic compounds like some pesticides, are widely believed to pass from the tissues of prey organisms into those of predators and to attain concentrations there which are several orders of magnitude higher than those in the tissues of the prey species. This phenomenon poses a specific threat to long-lived organisms at the higher trophic levels.

Studies of bioaccumulation are carried out in the laboratory, in experimental ecosystems and in the field. Laboratory investigations are usually concerned initially with determining the 'bioconcentration factor' (BCF), that is the ratio between the concentration in the animals and the concentration in the water, when the animals have been exposed for sufficiently long for an equilbrium or steady state to be achieved. This ratio is generally regarded as a valid indicator of the capacity of a pollutant to accumulate in animal tissues. Such limited data as is available (Davies & Dobbs, 1984; Schnoor, 1982) suggest that laboratory-derived BCF values agree reasonably well with those derived from field observations on the animals of polluted waters, at least for certain groups of organic pollutants and provided certain conditions are met in the laboratory determinations. Under certain conditions, there is a good correlation between log P (where P = the octanol-water partition coefficient of the chemical) and the BCF. This offers the possibility that the bioaccumulation potential of a pollutant can be indicated by the result of a relatively simple chemical determination, rather than by the expensive and time-consuming estimation of BCF. However, Davies & Dobbs (1984) in their study of this question, found it necessary to reject determinations of BCF which did not meet certain criteria.

There are many models of bioaccumulation. In the simplest possible model, two compartments are considered: the organism and the environment. Pollutant will enter the organism at a certain rate, which is dependent upon the amount present in the environment. Pollutant will also be lost from the organism, at a rate dependent upon the amount present in the organism. This simple model can be expressed mathematically and predicts that organisms exposed to a constant level of pollutant will eventually reach a 'steady state' that is the concentration of pollutant will increase to a certain level and thereafter remain constant. Conversely, the model predicts that in contaminated organisms maintained in clean water, the concentration of pollutant in the organism will decline exponentially. These predictions are generally confirmed by experimental findings.

Such a simple model is of limited practical use, however, and it is not difficult to see why. Most organisms cannot be considered as a single compartment. Studies of the distribution of pollutants in animals invariably show that the pollutant is very unevenly distributed between the various body tissues. Different pollutants behave in different ways. Clearly the animal will begin to suffer harm when the poison concentration in a particular organ reaches a critical level. Thus the concentration of the pollutant in the whole body is not a good indicator of harmful effect. For this reason, models have been derived which treat the organism as a set of interacting compartments. These models treat discrete organs (e.g. liver, kidney, brain) as interacting compartments connected via the blood (itself considered a compartment) with each other and with the external environment. Further, in some models the exchange of pollutants between the compartments is considered as a series of separate processes. For example, uptake of the pollutant from the environment may occur through the body surface, or by ingestion of food or of non-food particulate material. Elimination of the pollutant may occur through outwards diffusion, through renal or gastro-intestinal excretion, or by metabolic breakdown. Clearly models which attempt a complete and accurate description of bioaccumulation soon become extremely complex, and for that reason they are not treated in detail here. Hamelink (1977) and Moriarty (1984) provide good discussions of the strengths and weaknesses of bioconcentration models.

The purpose of such models is twofold. Firstly, if they can be validated by experimental findings (i.e. if the predictions of the models correspond with actual observations), they provide the means for making useful predictions based on comparatively simple experimental measurements. In other words, they can eventually become a substitute for actual experimentation, which may be time-consuming and expensive. Secondly, they can provide valuable information about the mechanisms of bioaccumulation. If, for example, an experimental finding does not agree with a prediction of a model, it indicates that one or more of the assumptions in the model is wrong, and thus focuses attention on areas which require further investigation. Although it is by no means clear that any existing model is of general application, studies on bioaccumulation and its mechanisms are of great practical importance. For instance, Moriarty (1984) has convincingly argued that the lack of understanding of bioaccumulation processes in single species has serious implications for ideas about the passage of pollutants through food chains.

It is widely believed that many pollutants pass through succeeding trophic levels and accumulate in high concentrations in the tissues of long-lived predators. That this is a general phenomenon, even for persistent pollutants like heavy metals and refractory organics, is in fact not well established. Moriarty (1984) has discussed some of the inadequacies of current knowledge. For example, comparing tissue levels of pollutants in field populations is very likely to produce biased results if, as is usually the case, mean levels of pollutants are compared. This is because of the differences in pollutant concentrations between individuals of the *same* species; frequently, mean values are biased by a very small number of individuals which have very high concentrations, that is the frequency distribution of pollutant concentration values is highly skewed. Further, although under experimental conditions a steady-state concentration of the pollutant in the tissues generally is eventually achieved, it is not clear that this is the case in the field. Field populations are generally exposed to lower, and more widely-fluctuating, pollutant concentrations in their environment. Clearly a comparison between tissue pollutant concentrations in two different species is invalid if they are not both at their respective steady-state concentrations. It is also true that the interpretation of field observations, and the design of experimental investigations, often rests upon unverified, and unwarranted, assumptions about what animals actually eat. Obviously in a simplified experimental system, predators will feed on prey which may not form their normal or natural diet in the field. A further example is the widespread assumption that large marine predatory fish such as tunas feed primarily or exclusively on smaller fish such as mackerel, herring or sardine. However, the stomach contents of several hundred tunas caught in the Mediterranean during an international angling contest consisted almost entirely of crustaceans and plankton (Fowler *et al.*, 1979). Thus while the question of bioaccumulation and biomagnification along food chains is an important and interesting one, authoritative answers require the solution of several technical and conceptual problems, and a greater knowledge of the basic ecology and physiology of many species.

5

Water pollution and public health

It is widely but erroneously believed that the marked increase in life expectancy which has occurred among the populations of many countries since the end of the nineteenth century was due to advances in medical science. In fact, in the conditions which prevailed in the swelling and poorly-sanitated cities of the nineteenth century the major hazards to life were epidemic diseases which exacted a heavy toll, particularly among the young. Certainly medical scientists played a major role in establishing the nature and the means of transmission of these diseases. Nevertheless the control of epidemic diseases was achieved largely through the development of proper methods for the treatment and disposal of wastes, and by attention to the provision of clean water supplies, together with education and legislation concerned with general hygiene. These developments preceded by some decades the availability of medical treatments, such as antibiotics, for the cure of waterborne infections. Today, those of us who live in the better-developed parts of the world can be reasonably optimistic, should we catch a waterborne disease, of a successful outcome, since most of these diseases respond well to modern medical treatments. Nevertheless we should not forget that the overwhelming majority of us will never become infected in the first place; and the reason for this is primarily the existence of adequate measures for the monitoring and control of pollution.

All this is of more than historical interest, since in many parts of the world today waterborne diseases remain a major hazard. They are endemic in those countries which have not yet established systems for the sanitary disposal of wastes. It is striking that in times of war or natural catastrophe, when sanitary systems cannot be maintained satisfactorily, waterborne diseases take very little time to spread through the human population. Even in countries where waterborne diseases are not considered endemic, the speed and frequency of international travel and the magnitude of international trade present a constant threat of the reintroduction and spread of infections. Further, increasing demands upon the water resources even of developed countries present new public health problems. Britain, for example, is not a country which could be considered as suffering from an overall shortage of water. Nevertheless the demands on the country's water resources are such that it is increasingly necessary for water to be re-used. The preferred sources of water supply are upland surface waters or ground water from deep wells or boreholes; these are unlikely to be contaminated with noxious chemicals or disease-causing organisms, and can be rendered suitable for potable water supply with minimal treatment. However, Evans & Johnson (1984) estimate that already 30% of Britain's water supply is derived from lowland rivers. Such water is likely already to have been used several times over by communities upstream, and to have received waste discharges from domestic, agricultural and industrial sources. The re-use of water poses special problems in relation to the spread of infectious diseases and of other harmful effects caused by chemical contaminants.

Public health measures for the control of infectious and other diseases associated with water take various forms. Examples include good medical services for rapid diagnosis and treatment; a system of rapid reporting of infectious diseases so that epidemics and their sources can be quickly identified and eliminated; education and general public awareness of good hygienic practices; and controls over the processing and handling of food for human consumption. The monitoring and control of water pollution is a central element in the preservation of public health, because water is potentially the means by which many diseases can be spread.

5.1 WATER POLLUTION AND PATHOGENS

Pathogenic organisms which are spread by polluted water include bacteria, viruses and parasites. Some common examples are discussed here, but it is important to remember that many infections are never identified as caused by any of the well-known agents of epidemic disease, particularly in the less well-developed areas of the world. In fact diarrhoea, which may be caused by bacterial, viral or parasitic infections is responsible, according to some estimates, for six million deaths per year (Slade & Ford, 1983). Some common means of transmission of enteric diseases are summarised in Fig. 5.1.

5.1.1 Bacterial pathogens

Typhoid fever is a disease of the gastrointestinal tract which frequently gives rise to systemic infections. If diagnosed and treated with antibiotics, it is debilitating but rarely fatal. Without treatment it is fatal in between 12% and 30% of cases (Hornick, 1982). The disease is rare wherever public health and pollution control measures are adequate, but minor outbreaks are not uncommon, particularly among travellers recently returned from less-developed parts of the world. Typhoid infection is usually caused by ingestion of bacteria from faecally-contaminated water or food. In Britain, the most serious epidemic of recent times affected over 500 patients. It was caused by canned meat imported from Argentina. It was discovered that the canned and sterilised meat had been cooled in contaminated water, and that water had leaked through imperfect seals in the cans (Ash *et al.*, 1964). The causative organism is *Salmonella typhi*. If the bacteria survive passage through the acid of the stomach (a process which may be assisted if the bacteria are ingested with water), they colonise the intestine and enter the epithelial cells of the gut lining. Very few bacteria (between 100 and 1000 cells) are required to establish a *Salmonella typhi* infection; for other *Salmonella* infections, the minimum infective dose is rather higher,

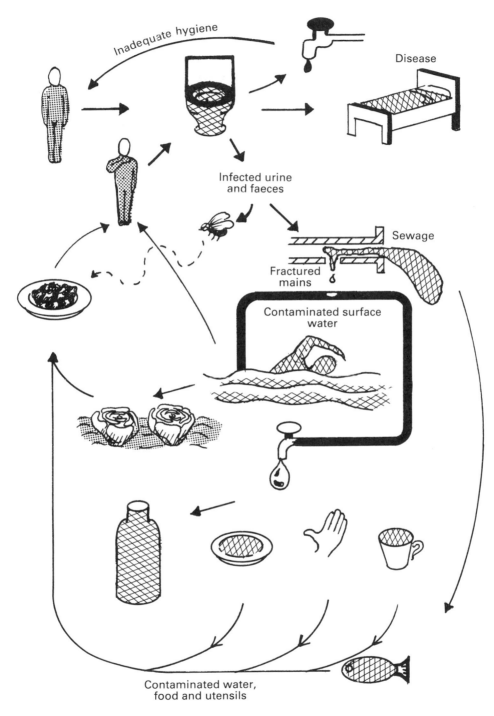

Inadequate hygiene

Disease

Infected urine
and faeces

Sewage

Fractured
mains

Contaminated surface
water

Contaminated water,
food and utensils

Fig. 5.1 — Summary of the modes of transmission of enteric diseases through contaminated
water.

approximately 10^5 to 10^6 cells. Ulceration of the intestine occurs, and bacteria enter the bloodstream. At this stage, fever occurs and the patient becomes mentally and physically debilitated. Subsequently, the bacteria may become established in various parts of the body, especially in the lymph nodes, gall bladder, spleen and skin — characteristic haemorrhagic spots on the skin may be seen in some patients. Infected individuals excrete large numbers of bacteria in the faeces, providing a source of infection of further individuals. An interesting and dangerous feature of this disease is that in some cases the symptoms of infection can be mild and undiagnosed, although the patient becomes chronically infected with *Salmonella* without showing symptoms of the disease. These 'carriers' are a persistent source of infection in the community; further, even if diagnosed the disease in such patients does not readily respond to the normal treatment with antibiotics. It is estimated (Hornick, 1982) that in the USA there are about three or four typhoid carriers per 100 000 in those populations which have been studied, although this number seems to be declining. However, it must be assumed that typhoid carriers will be present in any sizeable community.

Bacterial dysentery (shigellosis) and its cause were first recognised in Japan at the end of the nineteenth century during an epidemic involving 90 000 cases. It was almost certainly a scourge of the human population long before its nature was understood. Keusch (1982) refers to descriptions of the disease in the Old Testament, and by the ancient Greek writer Thucydides. Today the disease remains common throughout the world. In the USA as recently as 1978, data from the Centre for Disease Control indicated nearly 25 cases per 100 000 among infants (Keusch, 1982). In a rural community in Guatemala during the period 1965–1969, the incidence rate among infants was almost 200 000 per 100 000 per year, that is each infant was attacked by the disease on average twice per year (Mata, 1978). Shigellosis is caused by the consumption, in faecally-contaminated water or food, of live bacteria of the genus *Shigella*; known pathogenic species include *Sh. dysenteriae*, *Sh. flexneri* and *Sh. sonnei*. All of these appear to be associated exclusively with humans and some non-human primates, and transmission of the disease is usually by the faecal-oral route. *Shigella sonnei* outbreaks, although less serious than *S. dysenteriae*, are not uncommon in schools and similar establishments. If ingested bacteria survive passage through the stomach, they invade the epithelial cells of the intestine and give rise to ulcerous lesions. It is believed that the bacteria also secrete a toxin which may have pathological effects. Infected patients excrete bacteria with the faeces. Like typhoid, the disease typically takes a few days to manifest itself. The classical symptoms are the frequent passage, in small volume, of stools accompanied by blood and mucus, along with severe abdominal cramps; often, however, watery diarrhoea is the only manifestation and many infected patients may be undiagnosed. Shigellosis is often a seriously debilitating disease which can give rise to chronic infection and periodic recurrence of symptoms. If untreated, death occurs in up to 15% of cases, most fatalities occurring among infants and elderly patients. Complications include haemolytic anaemia and Reiter's syndrome, an arthritic condition. As with typhoid fever, chronically-infected patients can act as carriers of the disease within the community without manifesting clinical symptoms of the disease.

Cholera is perhaps the most devastating of the waterborne bacterial diseases, and well exemplifies the importance to public health of good sanitation and water pollution control. Cholera epidemics are by no means uncommon, especially in countries where adequate waste disposal practices are not well established. Benenson (1982) in his account of the disease, considered that it was possible to identify seven cholera pandemics; the most recent of these he considered to have originated in Indonesia in 1958 and to have spread to South-West Asia, India, the Middle East, Africa and to some countries in Southern Europe by 1976. At the time of writing Benenson (1982) considered this pandemic to be still extant. Earlier epidemics affected Britain and other European countries (Benenson, 1982; Fraser, 1984). The disease appears to be endemic to areas of Asia and to have first been recognised in Europe in the early sixteenth century, almost certainly introduced by a sailor returning from a foreign journey. The frequency of foreign travel in modern times presents a constant threat of the reintroduction and spread of the disease even in countries where good pollution control practices are established.

Cholera is caused by the bacterium *Vibrio cholerae*, and is transmitted by ingestion of live bacteria from polluted water or other material contaminated with faecal matter from an infected individual. The bacteria colonise the intestinal tract and produce a potent toxin which attacks the intestinal mucosa and interferes with the normal processes of salt and water balance which occur across the gut wall. The patient suffers severe diarrhoea involving sudden and massive dehydration of the body and serious salt imbalance which impairs the normal function of many organs. Death, due to circulatory failure consequent upon dehydration and salt imbalance, can occur within hours of infection. In some cases the bacterium can cause serious infections in other parts of the body.

Bacteria of the genus *Vibrio* have been studied mainly because of the medical importance of *V. cholerae*, which was first recognised about one hundred years ago. More recently, a wider biological perspective upon the genus *Vibrio* has become available, which is important when considering the status of these organisms in relation to water pollution (Colwell, 1984). Vibrios appear to be primarily free-living organisms of soil and water, particularly associated with saline environments such as estuaries and salt marshes. Apart from *V. cholerae*, several other *Vibrio* species are pathogenic to humans, and some are pathogenic in aquatic organisms including crustaceans and molluscs. These animals are primarily detritivores and filter-feeders, and frequently accumulate high concentrations of *Vibrio* bacteria in their bodies. This gives rise to a hazard of human infection, especially in areas where shellfish are cultured, harvested, processed or consumed in significant quantities.

5.1.2 Viral pathogens

At the time when the relationship between water pollution and the spread of bacterial diseases was first understood, the role of viruses was hardly recognised. Today, the importance of viruses as agents of disease is well known, and it is clear that the pollution of water with human wastes is a major potential source of serious disease.

The viruses of greatest interest in this context are the group known as enteric viruses. The taxonomy of viruses is complex and unfamiliar to many biologists, and is based not only on the morphological and physical characteristics of the organisms but also upon their chemical and immunological properties. The term 'enteric viruses' is one of convenience rather than a distinct taxonomic grouping. Enteric viruses include agents from widely different taxonomic groups, but they have in common the fact that the intestine is their primary lodgement site. As they are commonly found also in water, they can potentially be transmitted by ingestion of water or other matter contaminated with faecal waste from an infected individual. Human pathogenic viruses that are commonly found in polluted water are listed in Table 5.1.

Table 5.1 — Human viruses that may be present in polluted water. Figures in parentheses are the number of known serotypes. Viruses marked with an asterisk are known to be spread by contaminated water, the others are human viruses frequently found in contaminated water which are considered at least potentially hazardous. After Rao & Melnick (1986)

Virus	Disease caused
Enteroviruses:	
*Polio (3)	Paralysis, meningitis, fever
Echo (34)	Meningitis, respiratory disease, rash, fever, gastroenteritis
Coxsackie A (24)	Herpangina, respiratory disease, meningitis, fever, hand, foot and mouth disease
Coxsackie B (6)	Myocarditis, congenital heart anomalies, rash, fever, meningitis, respiratory disease, pleurodynia
New enteroviruses types 68 to 71 (4)	Meningitis, encephalitis, respiratory disease, rash, fever, acute haemorrhagic conjunctivitis
*Hepatitis A (enterovirus 72)	Infectious hepatitis
Norwalk (2)	Epidemic vomiting, diarrhoea, fever
Rotavirus (4)	Gastroenteritis, diarrhoea
Reovirus (3)	Unknown
Parvoviruses:	
Adeno-associated (3)	Unknown
Adenovirus (30)	Respiratory disease, conjunctivitis, gastroenteritis
Cytomegalovirus (1)	Infectious mononucleosis, hepatitis, pneumonitis, immunological deficiency
Papovavirus SV40-like (2)	Immunosuppression, progressive multi-focal leucoencephaly

However, the link water-patient-water-patient has so far been firmly established only for poliovirus and Hepatitis type A virus.

In many cases, viral infections caused by these agents do not cause serious symptoms, but acute gastrointestinal and diarrhoeal illnesses are the commonest waterborne diseases in the more developed countries, and they may have serious consequences in some patients, particularly in infants, elderly patients and immunologically compromised hosts. In addition, some enteric viruses are capable of causing seriously debilitating or even fatal illness, and can give rise to epidemics. Consequently they are considered as serious hazards to public health, and some examples are discussed briefly here. A fuller account of the subject is given in a concise form by Rao & Melnick (1986).

Poliovirus is the causative organism of poliomyelitis, probably the most familiar of the waterborne viral diseases. In developed countries, public health policy generally dictates that the majority of the population is immunised in childhood against this disease. Poliovirus infection is primarily one of the alimentary tract and may cause symptoms of fever, vomiting and diarrhoea. These symptoms may not be recognised as serious. In a minority of cases (up to approximately 1%) the virus enters the bloodstream of the infected individual and may penetrate the nervous system and give rise to paralysis; this may be permanent and, if vital organs are affected (such as respiratory muscles), fatal. Infected individuals, whether or not showing serious disease symptoms, excrete poliovirus in large quantities in the faeces, giving rise to the risk of infection of further individuals.

Hepatitis A (infectious hepatitis) is clearly transmitted through contaminated water, and numerous outbreaks of the disease have been traced to this source (Rao & Melnick, 1986; Vaughn & Landry, 1983). Viral hepatitis of the non-A, non-B form has also been shown to give rise to waterborne epidemics with fatality rates of up to 40% among susceptible groups of the population (Rao & Melnick, 1986). There is strong circumstantial evidence of the waterborne spread of rotavirus, Norwalk virus and of infections caused by consumption of shellfish from polluted water which contain high concentrations of parvovirus, rotavirus and small round virus (SRV).

5.1.3 Parasitic infections

Parasitic infections can also be spread through polluted water. Amoebic dysentery, caused by the protozoan *Entamoeba histolytica*, is a typical example. Its life cycle consists of four stages: trophozoite, pre-cyst, cyst and metacyst. Ingestion of the mature cyst from contaminated food or water initiates the infection. In the small intestine the cyst gives rise to the metacyst, which in turn divides to give a number of amoebae. These enter the large intestine, where they may invade the host tissues or give rise to further cysts. Infected patients excrete trophozoites (amoebae) and cysts in the faeces, though only the cysts can survive in the external environment and survive passage through the stomach to initiate a new infection. Within the large intestine, invasion of the host tissues causes ulceration and abscesses, which may spread to other organs such as the liver. The most obvious symptom of acute amoebic dysentery is diarrhoea, accompanied by blood and mucus. Chronic infection can occur giving rise to recurrent episodes of diarrhoea alternating with constipation. Other parasitic protozoans with relatively simple life cycles are known to spread in a similar manner. Examples include the flagellate *Giardia lamblia*, which causes serious gastrointestinal disorders; major outbreaks continue to be recorded in Europe and the USA (Zaman and Ah Keong, 1982).

Parasitic nematodes (roundworms) are important human parasites, and many have a simple life cycle involving direct transmission by the faecal-oral route. Others require passage through an intermediate host, but completion of the life cycle is facilitated by contamination of water, soil or food with faecal wastes of infected individuals. In many species, the life cycle includes a stage which is specifically adapted for survival in the external environment, so that infective life stages may remain viable in water or soil for long periods. In some species, a period in the external environment is essential for completion of the life cycle. One example of a nematode with a simple life cycle which causes human infection is the whipworm *Trichuris trichiura*. Mild infections produce few serious symptoms, but provide a focus of infection from which more serious cases can arise. Heavy infection produces symptoms of dysentery, and leads to severe damage of the intestinal tract, anaemia and (especially in children) the loss of weight. The faeces of infected individuals contain eggs which require about 20 days in the external environment to complete their development; the infective stage can survive for several months in water or soil. Several other nematode species have similar life cycles.

Tapeworms (cestodes) include a number of species which infect humans and whose transmission is aided by contamination of water or soil with human faecal matter. The beef tapeworm, *Taenia saginata*, is a typical example; again, there are a number of other species whose life cycles are broadly similar. The adult *T. saginata* lives in the human intestine and sheds, in the faeces of the host, proglottides which contain eggs. The eggs are ingested by cattle, from polluted water or from pasture contaminated with human faeces. In the cattle, the eggs develop through several stages, migrating through the intestinal wall and via the circulation to the muscles, where the cysticercus stage is formed. This encysted state remains in the muscle, and if ingested by a human in uncooked or partially-cooked beef, develops into an adult tapeworm.

Trematode parasites which infect humans include a number of blood flukes (Schistosomes), liver flukes, intestinal flukes and pulmonary flukes which cause serious disease. They are relatively rare in temperate zones and in countries where pollution control and public health practices are well established, but are major causes of ill health and economic loss in many parts of the world (Zaman & Ah Keong, 1982). Their life cycles are complex, involving one or more intermediate hosts, and all require a period of development within the body of a mollusc, usually an aquatic snail. Invariably, transmission of these parasites involves the contamination of water with human faeces. The life cycle of a typical blood fluke is shown in Fig. 5.2.

Control of the spread of parasitic diseases is an enormous public health problem in many parts of the world. Detailed consideration of the life cycles of parasites is required in order to understand their modes of transmission, the control measures which may be effective against them and the relationship between water pollution and the spread of parasitic infection. Where the life cycle includes one or more intermediate hosts, control measures may be directed against the intermediate host in order to break the cycle of infection. Aquatic molluscs living in polluted waters can readily become established as a reservoir of human infection. It is often important, therefore, to eliminate aquatic molluscs by various means, including the use of selective poisons. In this case, an understanding of the toxicological principles and techniques discussed in Chapter 4 clearly becomes important. Alternative methods of eliminating aquatic molluscs include management of the flow regime of drainage

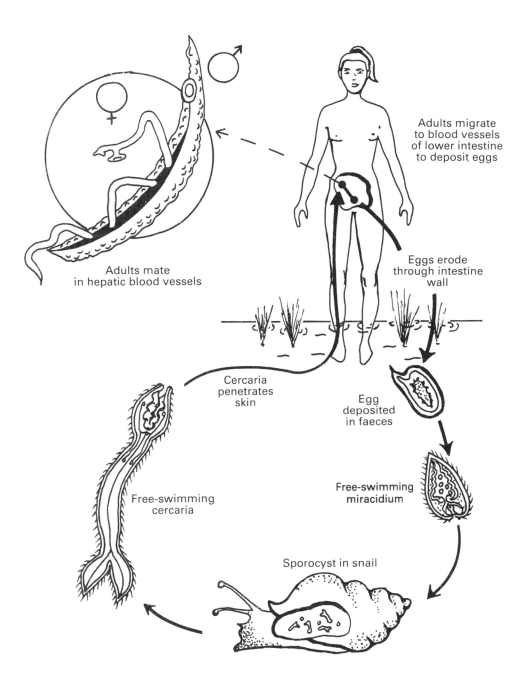

Adults migrate
to blood vessels
of lower intestine
to deposit eggs

Eggs erode
through intestine
wall

Adults mate
in hepatic blood vessels

Cercaria
penetrates
skin

Egg
deposited
in faeces

Free-swimming
cercaria

Free-swimming
miracidium

Sporocyst in snail

Fig. 5.2 — Life cycle of a typical trematode parasite of humans, *Schistosoma mansoni* (not to scale).

and irrigation channels, where this is technically feasible, since many aquatic molluscs are restricted to static or slow-flowing waters.

Strict regulation of food processing and handling is also important, including effective systems for the inspection of meat and other foodstuffs to avoid infected material being used for human consumption. Education of the general public in personal hygiene, and in the appropriate techniques for the preparation and cooking of food can do much to limit the spread of disease. Deep freezing food (−20°C for 24 h) is usually sufficient to kill infective stages of parasitic worms. Adequate medical facilities for the diagnosis and treatment of infection are of obvious importance. Nevertheless a large proportion of the most serious parasitic diseases — whether caused by protozoa, nematodes, cestodes or trematodes — have in common that their transmission is facilitated by the contamination of watercourses or land with human faecal wastes. The monitoring and control of water pollution, and the development of adequate methods for the safe treatment and disposal of wastes, is therefore arguably the most powerful single measure that can be taken against many parasitic diseases.

5.1.4   Pathogens in the aquatic environment and in wastewater treatment processes

Deprived of the favourable growth conditions provided by their hosts, many pathogens die quite quickly. They may, however, remain viable sufficiently long to create a hazard to the health of bathers or other recreational users of the receiving water, and to render the water unsuitable for domestic or agricultural uses. It is therefore necessary to consider the survival rate of pathogens in the aquatic environment, and the effectiveness of waste treatment processes in removing them from effluents.

Some representative data on the survival of bacteria are summarised by James H. Montgomery Inc. (1985). For a range of enteric bacteria and pathogens in well water, times for 50% reduction in the initial population at 9.5°C to 12.5°C varied from 2.4 to 24 hours. For a 99.9% reduction, however, much longer periods of 20 days or more appear to be necessary for some pathogens. Factors which influence the survival times of bacteria include the temperature, the light intensity and the quality of the water. It is therefore highly desirable that sewage treatment processes remove as many pathogenic bacteria as possible. Conventional treatment processes (see Chapter 6) are often reasonably effective in removing bacterial pathogens from the final effluent, although the pathogens may become concentrated in the sludge. Further, treatment plants cannot always be operated at maximum efficiency, particularly at times of high storm-water input. Under ideal conditions, primary and secondary treatment of sewage reduce the numbers of pathogens in the effluent by 90% or more, compared with the raw sewage. Much lower removal rates are obtained during the frequent periods when treatment plants do not operate under ideal conditions.

Viruses survive at least as long as bacteria. Sattar (1981) in his summary of a large number of reports refers to enteric viruses in natural fresh waters requiring from less than one day to over 21 days for 99% loss of infectivity. In sea water, there are several reports of periods over 100 days for 99% loss of infectivity. A wide range of physical, chemical and biological factors have been reported to influence virus survival (Sattar, 1981; Block, 1983). The behaviour of viruses in sewage treatment plants is almost impossible to predict (Gerba, 1981; Sorber, 1983). In some cases, almost complete removal of virus particles has been reported, whereas in others the viruses have been found to pass freely. It is possible that high removal rates can be achieved in pilot plants operating under continuously optimal conditions, but that under the varying and often sub-optimal conditions which prevail in operational treatment plants, virus removal is much less efficient. As with bacteria, it is likely that viruses removed from the liquid effluent are to a large extent retained in the sludge. While conventional primary and secondary treatment cannot be considered generally effective in virus removal, some forms of tertiary treatment have been found to be successful (Rao & Melnick, 1986). Tertiary treatment of sewage, however, is still carried out only exceptionally even in better-developed communities.

At least some parasitic organisms may be expected to survive in the environment for long periods. The life cycle of many parasites includes a stage which is specifically adapted for survival in the external environment, to facilitate passage from one host to another. Some intestinal nematode parasites of humans which do not require any intermediate host do require a period in the external environment in order to complete their life cycle. Being relatively large, they settle readily and are often efficiently removed from sewage effluent by conventional treatment processes, although again they may tend to be concentrated in the sludge.

5.2 MONITORING PATHOGENS IN WATER

Since practicable waste treatment processes, however efficient, do not remove all the pathogens from sewage it is necessary to monitor receiving waters for the presence of pathogens. This is particularly necessary in waters which are used for bathing, fisheries or as a source of further water supply. It also provides information on the operational efficiency of the treatment plant. Because there are so many potential pathogens, in routine monitoring generally no attempt is made to identify them all. Most commonly, the presence or level of abundance of coliform bacteria is used as an overall indicator of contamination of the water with faecal material. Coliform bacteria are so-called because they resemble *Escherichia coli*, a normal inhabitant of the human intestine which is excreted in vast numbers in the faeces of healthy individuals. *E. coli* itself is not normally pathogenic, although some other coliforms are. Depending upon the environmental conditions, coliforms survive only a few hours or days outside their hosts; therefore their presence in water suggests that it has been recently contaminated, and that pathogens are likely to be present. Further, because the non-pathogenic *E. coli* is excreted by healthy individuals, it will nearly always be far more abundant in sewage than other coliforms of other enteric pathogens. Therefore by ensuring that the numbers of coliforms are kept at a low level, it is possible to be reasonably certain that the numbers of pathogens present are very small indeed. Coliform bacteria can be readily isolated, identified and counted from water samples (see section 5.2.1).

Despite these advantages, coliforms are by no means the only organisms which can be used as indicators of faecal contamination. Not all coliforms are of human faecal origin, so simply counting the total number of coliforms present can be misleading. It also sometimes happens that pathogens may be present when coliforms are absent or rare. The different types of coliform can be distinguished from one another by biochemical tests, and it is often suggested that specifically human faecal coliforms, rather than total coliforms, are a better indicator of sewage contamination (Bonde, 1977). Other bacterial indicators are available, such as faecal streptococci; this group is also easy to identify and occurs in large numbers in the faeces of healthy individuals. Pathogenic viruses, and the infective life stages of parasites, behave in many respects differently from bacteria. Although the presence of viruses or parasites is not routinely used to indicate faecal contamination, there are circumstances where it is advisable, for reasons of public health, to monitor water bodies for their presence.

5.2.1 Monitoring for coliforms

Coliforms are Gram-negative rod-shaped bacteria which are most readily distinguished from others by means of their biochemical properties. A common test for the presence of coliforms in water is to inoculate a suitable liquid growth medium with a quantity of the water and to observe the result. The medium used (lactose broth) contains lactose, a sugar which is fermented by coliforms to produce acid and gas. The medium may also contain substances which inhibit the growth of non-coliform bacteria while promoting the growth of coliforms. The production of acid is detected by incorporating an indicator dye in the medium; gas production is detected by inserting into the culture vessel a small inverted test tube (Durham tube). After a period of incubation at 35°C, positive cultures are easily recognised by the fact that the medium has changed colour and a bubble of gas has collected at the top of the Durham tube. This simple presumptive test for coliforms does not distinguish between coliforms of faecal origin and those from other sources. Confirmation of the presence of faecal coliforms requires that positive cultures are tested on a solid medium, lactose-peptone agar containing sodium sulphite and basic fuchsine. On this medium, colonies of faecal coliforms acquire a greenish colour, non-faecal coliforms a red colour and non-coliform colonies are uncoloured. Alternatively, a definitive one-stage test for faecal coliforms is to test for acid and gas production in lactose broth at 44.5°C. However, this test gives misleading results unless the incubation temperature is very precisely controlled. The growth media required are readily available commercially in made-up form.

Estimation of the numbers of coliforms present in a sample of water is done in one of two ways, the plate-count method or the Most Probable Number (MPN) method. In the plate-count method, an aliquot of the test water, usually about 1 ml, is mixed with the growth medium before the agar has set. When the medium has solidified and the plate has been incubated, each viable bacterium in the original aliquot will have given rise to a colony on the plate which can be readily observed. If a number of replicate plates are set up, the mean number of coliforms per ml of water can be calculated, along with its confidence limits, giving a reasonably accurate estimate of the number of coliforms present in the water.

A variation of the plate count technique which has some practical advantages is to use commercially-available millipore filters with a pore size of 0.45 microns. A known volume of water is filtered through the sterile filter disc, which is subsequently placed on a suitable solid medium. The number of colonies which develop during incubation corresponds to the number of viable cells present in the original sample.

The MPN method is simpler but much less accurate. Suppose that the sample of water actually contains 100 bacteria per ml. A series of tenfold dilutions will each contain, on average, ten bacteria per ml. If each of these dilutions is further diluted by a factor of ten, and an aliquot of 1 ml taken from each, then on average each 1 ml aliquot will now contain one bacterium. However, some will contain more than one, and some will contain none. If these aliquots are incubated in liquid medium as described above, those aliquots which contain no bacteria will score negative, and those which contain one or more will score positive. If aliquots from the original tenfold dilution were incubated in a similar fashion, very few of them would be negative, since the probability of a 1 ml aliquot containing no bacteria is small when the average content of bacteria is 10 per ml. However if a 1000-fold dilution of the original sample were made, only one aliquot out of ten would be expected to contain a bacterium and to score positive. This line of reasoning forms the basis of the MPN method, since for a given concentration of bacteria in the original sample, a given dilution factor and a given aliquot size the ratio of positive to negative results that would be expected on average can be calculated statistically. In practice, there are a number of variants of the MPN procedure, but all are based on the same principle. One way to proceed is progressively to dilute the original sample by factors of 10. From each dilution, a number of culture tubes (usually three to five) are inoculated with an aliquot of the diluted sample and incubated. When the bacteria are present in high concentration, all five tubes will score positive. When the dilution is very great, all five tubes will be negative. At intermediate dilutions, some tubes will be positive and some will be negative.

In the example shown in Fig. 5.3, aliquots of the original sample have been inoculated into liquid media in groups of five tubes, so that the original sample is diluted to $10^{-1}, 10^{-2}, 10^{-3}, 10^{-4}, 10^{-5}$ and 10^{-6} of its original strength. As expected, at the highest dilutions (10^{-5} and 10^{-6}), all the tubes score negative — no coliforms were present in any of the inocula. At the lower dilutions ($10^{-1}, 10^{-2}$), all five are positive, that is each tube was inoculated with at least one coliform. At the intermediate dilutions (10^{-3} and 10^{-4}), some tubes are positive and some are negative. The three sets of tubes indicated in the diagram are the ones used to determine the MPN, the number of positive scores in each set of tubes being five, two and one respectively. Using a table of MPN values specifically calculated for this procedure, the MPN corresponding to a score of 5, 2, 1 can be determined. Note that, starting from the left of the diagram, scores of 5, 5, 2; 5, 2, 1; 2, 1, 0; or 1, 0, 0 could be generated. Every MPN protocol specifies a set of rules for deciding which sets of tubes are used to obtain the score. In this case, the rule is 'select the greatest dilution giving some positive reactions together with the two preceding sets of dilutions'. Numerous variants of the basic MPN procedure are employed. These variations include the number of tubes tested at each dilution, the range of dilutions required, the size of the inoculum and the method of scoring. They are necessary because the numbers of bacteria found in raw sewage, river water and treated

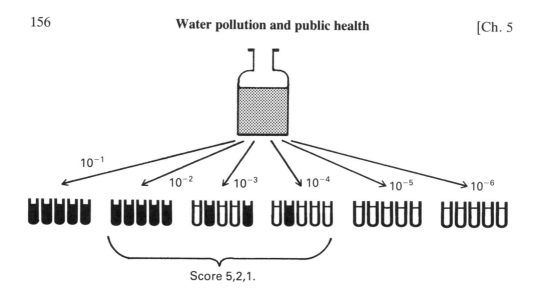

Fig. 5.3 — Determination of the most probable number of coliforms in a sample of water. See text for explanation.

drinking water are obviously greatly different. It is important to realise, therefore, that MPN tables can be used only in conjunction with the exact procedure for which they were calculated. Detailed descriptions of various MPN procedures for different purposes are given, for example, by the American Public Health Association (1975) and by HMSO (1983b).

An important disadvantage of the MPN procedure is that it is intrinsically inaccurate and potentially misleading. Assume that an MPN value is obtained of 100 organisms per 100 ml of water. This figure may be the most probable number, but a value of 80, or 120, is only slightly less probable; and the probability that the true number is either 80 or 120 could be greater than the probability that the true number is actually 100. In fact, although 100 is the most probable single number, it is almost certain that the true number is some greater or lesser value. Most MPN tables show, in addition to the MPN itself, the 95% confidence limits of the calculated value. Strictly, the test is incapable of estimating the number of bacteria in the original sample; it can only estimate with 95% confidence that the number of bacteria lies within a certain range. If this range were sufficiently narrow, this would not matter very much. In practice, however, the range is wide. For example, most published MPN tables show that for an MPN value of around 100 organisms per 100 ml, the 95% confidence limits embrace values from about 25 to about 270. Generally, the method cannot reliably distinguish between bacterial counts an order of magnitude apart. The MPN procedure is widely and routinely used in water quality monitoring, and arguably its limitations are not as fully appreciated as they should be.

Monitoring for other bacteria, and indeed for other micro-organisms such as yeasts, algae and protozoa can often be carried out (whether qualitatively or quantitatively) using procedures similar in principle to those used for coliforms, but with the appropriate culture media.

5.2.2 Monitoring for viruses

Isolating, identifying and counting viruses from the aquatic environment is intrinsically more difficult than monitoring bacteria, and there is less agreement on standard methods (Dobberkau *et al.*, 1981; Gerba, 1983; Rao & Melnick, 1986). As they are generally far less numerous than bacteria, and because the efficiency of virus recovery is often low, large volumes of water (100 litres or more) sometimes have to be processed by special filtration procedures and other complex techniques for concentrating the virus particles. The viruses must then be inoculated into cultures of living cells and allowed to grow. They are then isolated from the cell cultures and identified. The identification of viruses itself requires specialised techniques, including the observations of their effects on the cells in culture, electron microscopy and (increasingly) immunological techniques. Many viruses of potential interest are notoriously difficult to grow in cell culture, or grow only very slowly. It takes much longer, often weeks rather than days, to obtain the eventual results than is the case with bacteria, and accurate quantitative results can be obtained only with great difficulty. For these reasons, and because of the specialist skills and facilities required for virological work, routine monitoring for viruses is not practised very widely, but it may become more common as new methods are developed. In view of the difficulties of isolating and identifying viruses from water, it would be advantageous for routine monitoring purposes to use a single, readily-identifiable virus species as an indicator of the possible presence of pathogenic viruses. One possibility is to use attenuated poliovirus for this purpose. In many countries, children are immunised routinely against polio by means of an oral vaccine containing live but attenuated (non-pathogenic) poliovirus. Thus in any large community, there will always be some people excreting the attenuated virus, and its presence in water might be taken to indicate that enteric viruses, including pathogens, are likely to be present in the water.

5.2.3 Monitoring for parasites

Monitoring for parasites requires skilled personnel rather than specialised equipment. The most common method of identifying parasites is direct visual examination, under the microscope, of the water sample by someone trained to recognise the various life stages of parasitic organisms. The identification may be aided by certain staining procedures, and quantitative results can be obtained by using standard microscopical techniques of counting. Sometimes it is necessary to concentrate the suspended matter in a water sample of known volume, for example by centrifugation. Some parasitic protozoa, including *Entamoeba histolytica*, can be cultured in suitable growth media and treated rather like bacteria (Diamond, 1983). In areas where amoebic dysentery is endemic, a group of non-pathogenic intestinal amoebae (of the genus *Entamoeba* but distinguishable fron *E. histolytica*) are useful indicators of the possible presence of pathogenic protozoa rather as coliforms can be used as indicators of pathogenic bacteria.

5.3 WATER POLLUTION AND WATER SUPPLY

An important consequence of water pollution is that polluted water is generally less acceptable than clean water for most of the purposes to which water may be put. To take an obvious example, water carrying a heavy burden of potentially pathogenic organisms may be unsuitable for use as a source of potable water supply, or may only be rendered suitable by means of elaborate and expensive purification processes. These difficulties are exacerbated by the increasing need for water to be re-used as demands on water resources increase.

5.3.1   Potable water supply

The requirement that potable water be free from pathogenic organisms is obviously of prime importance, and water treatment processes are designed to achieve this end. Conventional water treatment processes are less effective in removing chemical contaminants which may give rise to harmful effects. Nitrates are common pollutants of water (see Chapter 2), and may be found in high concentrations in water bodies used as sources for potable supply. Excessive nitrate in drinking water is dangerous, particularly to young children and to babies fed on dried or bottled milk. Nitrate ions are readily converted in the stomach (at the low pH of gastric juices) to nitrite ions which pass readily into the bloodstream. The nitrite ions react with the haemoglobin of the blood, forming methaemoglobin, a relatively stable compound which has a lower oxygen-carrying capacity than haemoglobin. Cyanosis (blueing of the skin) is the most obvious symptom of severe methaemoglobinaemia. Some thousands of fatalities have been recorded around the world, and in many countries nitrate in water supplies is routinely monitored to ensure that the concentration does not exceed 50–100 mg l^{-1}. In some areas of Britain, for example, it is necessary to distribute bottled water at times when the nitrate levels in the public water supply exceed the safe level. Nitrates (converted to nitrites in the stomach) can also react in the gut with amines to form potent carcinogens, although the epidemiological significance of this process is not yet clear. Benes (1978) lists, in addition to nitrates, a large number of potentially toxic substances which can occur, in trace amounts, in drinking water derived from polluted sources. These include trace metals, pesticides, and a large number of organic compounds, some of which are derived from industrial or agricultural pollution, some from leaching of chemicals from plastic and other materials used in the water distribution system, and even some derived from chemicals used in the processing of water and wastewater. The significance of these 'micropollutants' is not well established, but their presence in drinking water is a potential cause for concern.

5.3.2   Agricultural water supply

Agricultural activities are a major source of serious water pollution, but at the same time agriculture imposes a heavy demand for supplies of clean water. Noy & Feinmesser (1977) briefly summarised some of the difficulties which can arise from the use of polluted water for agricultural purposes. The seriousness of such problems depends, to a large extent, on the varying conditions which prevail in different parts of the world, especially in relation to the climate, the availability of water and the extent to which waste treatment facilities and general public health provisions are established. The volume edited by Shuval (1977a) includes examples from several different countries. The disposal of wastewaters, whether treated or not, and of sludges from treatment processes to agricultural land has many economic and agricultural benefits, but there are a number of hazards associated with these practices. Even where the deliberate re-use of water is not practised, it is increasingly likely that agricultural water supplies will be drawn from water bodies which have previously received waste matter.

The most obvious adverse effects of polluted water in agriculture relate to the presence of toxic matter, especially heavy metals, and of pathogenic organisms. Agricultural animals probably do not differ much from humans in their sensitivity to toxic heavy metals, and heavily-contaminated water is no more acceptable to them than it would be as a potable supply for humans. The sensitivity of plant species to heavy metals in their environment is well known. Although there is some possibility that crops may accumulate sufficient heavy metal to be hazardous to consumers, in practice the effects of toxic metals are initially economic; some crops will give a reduced yield, or fail altogether, if the levels of toxic metals in the soil or irrigation water are too high. Thus water pollution can restrict the uses to which land can usefully be put, or impose extra costs relating to the supply of water of adequate quality. Boron, for example, is like many other elements an essential requirement in trace amounts for plants; in excess, however, it is toxic. The widespread use of perborates in detergent formulations has led to concern that domestic sewage effluent, without any contamination from industrial sources, could so elevate the boron content of receiving waters as to exert an adverse effect on crops. Consequently it has been necessary for national and international agricultural agencies to formulate detailed recommendations concerning the chemical quality of water used for various agricultural purposes.

Pathogenic organisms in water or sludges applied to agricultural land present obvious potential dangers. Again, the magnitude of the hazard depends greatly upon the climatic, agricultural and general public health situation of the geographical area concerned. Infective life stages of parasites can be ingested by animals from contaminated water and transmitted to human consumers; some parasites may be transmitted directly to human consumers, particularly from salad vegetables or other crops which are eaten uncooked. The hazards from pathogenic bacteria and viruses have not been properly evaluated in many parts of the world, but Shuval (1977b) and Vaughn & Landry (1983) briefly reviewed some studies which suggest that microbial contamination of arable crops through irrigation with contaminated water can contribute to the spread of disease.

5.3.3 Industrial water supply

Industrial usage accounts for a very large proportion of the total demand for water, and about two-thirds of the water used in industrial processes is used for cooling. The use in some industrial processes of sewage or other effluents, or of water which is too polluted for other water supply purposes, would appear to offer some advantages. In practice, the use of polluted water in industrial processes is subject to certain constraints. For example, the dairy and food-processing industries obviously require water of the highest quality, and it is not always economically feasible to construct and safely operate water distribution systems in which water of different quality is supplied to different factories within an area. Many of the problems which can arise are not of a strictly biological nature; for example, the presence of ammonia and other chemicals can exacerbate corrosion problems in pipework. A common difficulty is the growth of bacterial slimes and the accumulation of organic detritus in cooling systems; this can be controlled by chlorination, provided that the presence of free chlorine at relatively high levels does not interfere with the normal operation of the factory. Shuval (1977b) drew attention to the possibility that the use of polluted water in cooling towers may spread pathogenic micro-organisms over a wide area. Eden *et al.* (1977) discuss water re-use practices in Britain, and describe some examples of the use of polluted water for industrial water supply. One pressure which is likely to increase in the future is the demand from some very modern industries, such as the manufacture of electronic components, for water of a very high standard of purity. This creates the need for manufacturing plants, often dealing with highly toxic chemicals, to be sited in the vicinity of the least-polluted water bodies of greatest utility and conservation value. Some biologists may take the view that such developments should be vigorously opposed. Others may argue that the application of known principles and practices and of others yet to be discovered will allow an optimum water-use strategy to be devised which would not necessarily prohibit proposals which are of considerable social and economic value. The socio-economic dimension is here not directly relevant; but we may be certain that as one biological problem is solved, another will surely arise.

6

Water pollution control

Biologists have an important function in the control of water pollution. Firstly, many waste treatment processes depend upon the controlled application of biological phenomena. Secondly, the assessment of the effects of pollution is ultimately a judgement which can only be made by biologists. Appropriate strategies for the treatment and disposal of wastes, and the assessment of the efficacy of such strategies, are therefore impossible without sound biological advice. Additionally, however, effective pollution control requires the expenditure of money, the professional advice of other scientific specialists such as chemists, engineers and hydrologists, and the formulation and enforcement of laws, regulations and administrative practices. The extent to which pollution is successfully controlled therefore depends upon constructive interactions between specialists from a diverse spectrum of disciplines.

The amount of money spent on the design, construction and operation of a sewage treatment works can be fairly precisely estimated. Any competent chemist can accurately determine the concentration of zinc, phenol or ammonia in a sample of water. Biologists, however, do have difficulty in estimating the number of fish or mayfly nymphs in a lake; and non-biologists may be forgiven if they do not easily understand why a million mayfly nymphs are preferable to an equal number of beetle larvae, particularly if they have been informed that in other places and under other circumstances the beetle larvae must be preserved! A legislative assembly may understand that while it can pass a law stating that all rivers must contain salmon, it is futile so to do; but the legislators may wonder why a concentration of ammonia which is harmless in one river may be devastating in another. The role of legislation is, however, clear. Since pollution control is expensive, technical advances do not necessarily lead to more effective pollution control in the absence of legislative, administrative and economic pressures which can be brought to bear on the individuals, industries and communities which are responsible for pollution. Equally, laws or international agreements on the control of pollution cannot be effective in the absence of the necessary technical and economic resources, including the availability of suitably skilled personnel.

Effective pollution control therefore depends upon the availability of adequate technical means, of an appropriate legislative and administrative framework, and of trained personnel drawn from a variety of disciplines who are able to communicate effectively with one another. In this chapter, some technical and administrative aspects of water pollution control are considered together.

6.1 BIOLOGICAL TREATMENT OF WASTEWATER

Sewage, and wastes consisting largely of putrescible organic matter, can be very effectively treated by biologically-based processes. In practice, these processes are often effective in treating wastes containing toxic substances, provided that the concentration of toxic matter is not too high. If untreated wastes are discharged to water they are broken down and assimilated into the receiving water ecosystem by a variety of physical, chemical and biological mechanisms. This may, however, cause disturbances in the receiving water ecosystem such as were described in Chapter 2. The objective of wastewater treatment is therefore to minimise these disturbances by ensuring that the processes responsible for the breakdown of organic matter, and for the removal of suspended solids, pathogens and toxic substances take place, as far as possible, within the treatment plant itself rather than in the receiving water. In this way the effluent will have a relatively small effect on the receiving water when it is discharged.

Biological waste treatment processes have evolved over a period of more than one hundred years, partly through trial and error but increasingly by design, as understanding of the underlying biological and engineering principles has improved. The biological aspects of waste treatment processes have been described and discussed by Mudrack & Kunst (1986), and in the volumes edited by Curds & Hawkes (1975, 1983a, 1983b). Numerous variations of the basic methods are available, but it is convenient to describe conventional processes of biological waste treatment in terms of a division into primary, secondary and tertiary stages. A typical treatment process can be summarised as in Fig. 6.1. The efficiency of the various stages of sewage treatment varies widely, according to the quality of the wastewater, the design of the plant and the operating conditions, which themselves vary widely, but some typical data are summarised in Table 6.1. Some other relevant data were given in Chapter 2 (Tables 2.1 and 2.2).

6.1.1 Primary treatment

The initial treatment of the wastewater is purely physical, consisting typically of screening and settling. The water is passed through series of screens formed from metal bars, to remove large objects such as bottles or pieces of wood. These large objects may be incinerated or otherwise disposed of, or may be comminuted into small particles and returned into the system. The water is then passed through one or more sedimentation tanks which are designed to allow a high proportion of the suspended particulate matter to settle at the bottom of the tank. The settled particulate matter (primary sludge) may be disposed of in various ways (see section 6.1.4), and the remaining liquid may then pass to the secondary treatment stage.

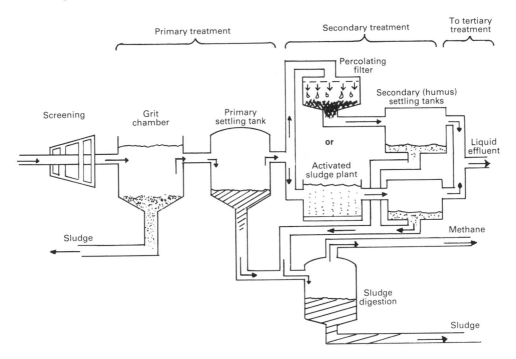

Fig. 6.1 — The different stages of a typical sewage treatment process.

Table 6.1 — Effect of various sewage treatment processes on the 5-day BOD, suspended solids content and bacterial numbers in sewage. The values given are those typically obtained under good operating conditions. Data from Klein (1966)

Process	Approximate percentage reduction (compared to values in raw sewage)		
	5-day BOD	Suspended solids	Bacteria
Sedimentation	30–40	40–75	25–75
Sedimentation + percolating filters	80–90	80–90	90–95
Sedimentation + activated sludge	85–95	85–95	90–98
Oxidation ponds	75–95	*a*	>99

a Suspended solids content of oxidation pond effluent can be very high, owing to the production of algae in the pond.

6.1.2 Secondary treatment

Secondary treatment, if applied, is most commonly undertaken by one of three methods: percolating filters, the activated sludge process, or oxidation (stabilisation) ponds. These are all essentially biologically-based processes in which organic matter is broken down aerobically by the metabolic activity of micro-organisms. It is important that the breakdown of organic matter occurs aerobically; under these conditions, organic compounds (mainly composed of the elements carbon, hydrogen, oxygen, phosphorus, nitrogen and sulphur) are rendered into relatively innocuous inorganic compounds such as carbon dioxide, water, carbonates, nitrates, sulphates and phosphates. Under anaerobic conditions, organic matter is broken down by different metabolic pathways (usually by predominantly different micro-organisms) to produce methane, ammonia, amines, organic compounds of phosphorus and hydrogen sulphide. These compounds are, in general, toxic to most forms of life as well as being aesthetically unacceptable. (Some of the reactions involved in aerobic and anaerobic decomposition processes were summarised in section 2.3.) Anaerobic conditions in waste treatment plants are therefore generally avoided, as they impair the efficiency of the treatment process and produce an effluent with undesirable and potentially harmful characteristics. There are, however, some circumstances — such as sludge treatment (section 6.1.4) — where the controlled use of anaerobic decomposition is beneficial. Also, anaerobic decomposition plays an important part in the functioning of oxidation ponds (see below).

Percolating (trickling) filters consist of circular or rectangular beds, usually about 2 m deep and from a few square metres up to about 2000 square metres in surface area, of stones, slag or synthetic material through which the settled sewage is allowed to percolate slowly (Fig. 6.2). The preferred material for constructing the filter bed is

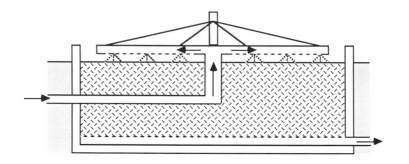

Fig. 6.2 — Cross-section through a percolating filter.

fairly coarse (25–50 mm) and of more or less uniform size to ensure that air-filled void spaces are plentiful. Materials with rough surfaces (high surface area/volume ratio) are advantageous. When sewage is allowed to trickle through the filter, the surfaces of the filter medium become colonised with a thin film of bacteria and other micro-organisms. This biological film is responsible for the breakdown of organic matter in the sewage. It typically contains heterotrophic bacteria which utilise complex organic compounds, and autotrophic bacteria which oxidise simple compounds, for example converting ammonia to nitrate. In addition, however, percolating filters support a variety of other organisms, including fungi, algae, protozoa, and many invertebrates, especially annelids, insect larvae, nematodes and rotifers. The relationships between these various organisms in the percolating filter, and their role in the treatment process, has been the subject of much study (Hawkes, 1963; Curds & Hawkes, 1975). Some appear simply to colonise, opportunistically, the favourable environment which the filter provides. Some of the filter biota are undoubtedly a potential source of nuisance; flies, for example, can become very numerous in the vicinity of trickling filters, and excessive growth of fungal mycelium within the filter can fill the void spaces, reducing the filter's efficiency and in severe cases rendering it inoperative. However, many of the organisms typically found in percolating filters make a positive contribution to its function. Protozoa feed on bacteria, and/or particulate organic matter, thereby contributing to the biological processing of the matter present in the sewage. The invertebrate species feed upon the meiofauna, on bacteria, and on particulate organic matter. They also browse upon the microbial film which is being continuously generated upon the surfaces of the filter medium. Their activities have three beneficial effects. Firstly, they prevent the microbial film becoming too thick; excessive growth of the film would eventually tend to reduce the efficiency of the filter. Secondly, they process organic matter — whether newly-synthesised by microbial activity, or that originally present in the sewage — consolidating it into faecal pellets which settle out more readily at the secondary sedimentation stage. Finally, through their respiration they are responsible for the conversion of a significant proportion of the organic matter to carbon dioxide and water.

In effect, the trickling filter is an intensely-productive semi-controlled ecosystem whose energy source is the organic material present in the sewage. The precise characteristics of the community which develops in the filter depend upon the quality of the sewage and the operating conditions of the filter. The qualitative and quantitative characteristics of the filter bed community can sometimes be used to assess the performance of the treatment process in much the same way that indicator organisms can be used to assess the biological quality of receiving waters (Chapter 3). Overall, the filter ensures that the organic matter present in the sewage is biologically utilised under aerobic conditions. Were this matter to be discharged untreated into the receiving water, its utilisation would result in deoxygenation of the water, with the undesirable consequences described in Chapter 2. The effluent from the percolating filter is very different from the settled sewage which entered it a few hours earlier (Table 6.1), although it still contains a lot of organic matter. A large proportion of this, however, is recently-elaborated biomass. The biological film and its associated grazing organisms gradually break off from the underlying filter medium, and the liquid effluent from the percolating filter therefore requires further settling in tanks similar to those used in primary treatment. This gives rise to a secondarily-treated liquid sewage effluent (which may be discharged, or subjected to tertiary treatment) and to secondary sludge, or humus. The secondary sludge has a high organic content, and may be disposed of by various means (section 6.1.4).

An alternative to percolating filters as a means of secondary treatment of sewage is the activated sludge process. In theory, secondary treatment by biological oxidation could be achieved simply by holding settled sewage in a suitable container and subjecting it to aeration. In practice, early experiments showed that the period of time required for acceptable treatment to be achieved by this simple approach was far too long. In addition, the exigencies of sewage treatment dictate that the process should be a continuous-flow rather than a batch-type operation. Various solutions to these problems have been devised, and the development and principles of the activated sludge process are described by Hawkes (1983a). The first operational plant which worked on a continuous flow basis was installed in Worcester, England, in 1916. The method is about equally as effective as percolating filters, but in some circumstances has economic and engineering advantages. It is cheaper to instal and requires less land, and nuisances caused by odours and insects associated with percolating filters are less troublesome. However, the process is more expensive to operate, and requires much stricter supervision of its operating conditions, so its suitability for a given location depends on local circumstances and the availability of qualified staff to operate the system.

There are many variants of the activated sludge process, but its basic principles are illustrated in Fig. 6.3. Settled sewage is introduced to a tank at a rate which is

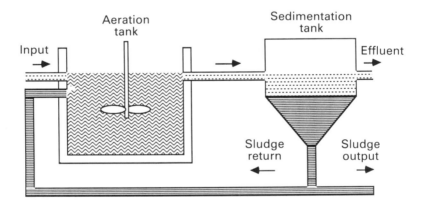

Fig. 6.3 — The activated sludge process.

designed to allow complete replacement of the tank's contents within a period of a few hours. During this time, the sewage in the tank is aerated vigorously. Microbial metabolism brings about breakdown of the organic material present in the sewage. As in the percolating filter, the end products of this breakdown are carbon dioxide, water, simple inorganic compounds and a considerable quantity of newly-synthesised organic matter consisting largely of microbial biomass. The effluent from the tank is passed to settling tanks in which the organic solids are separated from the liquid component of the effluent. However, an essential feature of the activated sludge process is that a small proportion of the effluent, including the organic solids, is returned directly to the reaction tank (Fig. 6.3).

The biological processes which occcur within the reaction tank have been the subject of much study (Hawkes, 1983a). Growth of bacteria, fungi, protozoa and microinvertebrates (particularly rotifers) results in the production of an organic floc. The physical and biological characteristics of this floc appear to be extremely important. The characteristics of the floc are greatly influenced by the operating conditions of the process, and in turn the efficiency of the process is clearly influenced by the characteristics of the floc. A plant which is performing badly is frequently found to have a floc whose physical appearance and biological composition is very different from that found in a plant which is operating well. In fact, regular inspection of the floc characteristics is widely used in monitoring the correct operation of the activated sludge process. A rather narrower range of organisms is typically found in activated sludge tanks than in percolating filters. Again, the question arises of whether these organisms are fortuitously colonising a favourable environment, or whether they have a specific and desirable role in the process. Experimental studies (Curds, 1975) have shown clearly that protozoa are responsible for a marked reduction in the turbidity of the effluent, and for the reduction in the numbers of bacteria present in the effluent. Effluents from plants containing protozoa also have lower values of BOD, COD, organic nitrogen and suspended solids compared to those from plants operating without protozoa. Rotifers also appear to be important. Doohan (1975) suggests that they are responsible for removing non-flocculated bacteria. They also promote floc formation, by producing faecal pellets and by breaking up large floc particles, both processes providing nuclei for the formation of further floc.

Oxidation (stabilisation) ponds are also widely used as a means of sewage treatment. Strictly speaking, oxidation ponds are not a means of secondary treatment, but rather an alternative means of complete sewage treatment to what have come to be regarded as conventional methods. Where oxidation ponds are employed, there is frequently no division of the treatment process into primary, secondary and tertiary. Some forms of tertiary treatment, as conventionally practised, in fact involve the treatment of secondary sewage effluent in what amounts to an oxidation pond (see section 6.1.3). As Hawkes (1983b) points out, treatment of used waters in biological oxidation processes involves the speeding-up and intensification of the natural processes of purification which occur in natural ecosystems. Oxidation ponds represent the lowest practicable level of this intensification, while percolating filters and activated sludge processes represent respectively higher levels. Thus to treat the sewage from 1000 people typically requires an area of 35 square metres in an activated sludge plant; but 210 square metres of percolating filter and between 2000 and 50 000 square metres of oxidation pond, depending upon climatic conditions, are required to treat the same amount of sewage to a similar standard. Oxidation ponds are therefore particularly associated with warmer climates and regions which are technologically less well-developed, although they are utilised in all parts of the world where local conditions are appropriate.

As with other sewage treatment processes, many variations of the basic procedure have been developed. A typical process is that described by Hawkes (1983b) as the facultative oxidation pond. A typical pond (Fig. 6.4) is about 1 m in depth. Waste is introduced near the bottom of the pond, which has a retention time of a few hours or days, depending upon local conditions. Frequently, ponds are arranged in

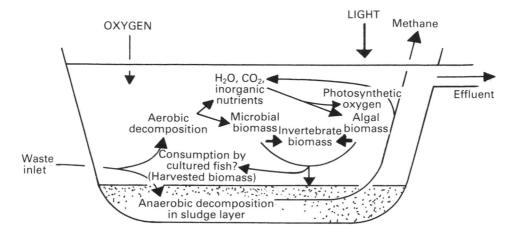

Fig. 6.4 — Summary of the processes which occur in a facultative oxidation pond.

series so that the effluent from one pond is passed to another for further treatment. The pond acts as a sedimentation tank, so that a layer of organically-rich sludge accumulates on the bottom. Within this sludge layer, anaerobic conditions quickly become established and decomposition of organic material takes place, resulting in the production of methane, carbon dioxide, ammonia and other metabolites typical of decomposition by anaerobic bacteria. These products themselves undergo further transformations in the upper, aerobic layers of the pond. Algae are particularly important in the functioning of oxidation ponds. If light is available, algal photosynthesis removes carbon dioxide and plant nutrients from the water, while providing photosynthetic oxygen. This allows the development of an aerobic bacterial flora which brings about further degradation of organic matter in the sewage. The waste material is thus broken down and resynthesised into new biomass. In part, this biomass eventually settles into the sludge layer and is recycled within the system. However, the newly-synthesised material provides a rich source of energy which is capable of supporting a complex ecosystem. In some cases, waste treatment can be combined with aquaculture so that useful biomass (e.g. plant material, fish) can actually be harvested from the system. As a means of sewage treatment, stabilisation ponds compare well, under good operating conditions, with the standards achieved by other processes (Hawkes, 1983b).

6.1.3   Tertiary treatment
After secondary treatment and sedimentation, the effluent is often discharged to the receiving water without further treatment. Tertiary treatment, if necessary, is applied to the effluent to bring about specific improvements to the secondary effluent before it is discharged. The precise nature of the tertiary treatment required depends upon the efficiency of the earlier stages of treatment, the characteristics of the receiving water and the uses to which the receiving water may subsequently be put. It is not therefore possible to describe a typical procedure for tertiary treatment, but

the following examples illustrate the range of treatments which may be employed. The principal reasons for applying tertiary treatment are to bring about a further reduction in the BOD, suspended solids, ammonia, nitrate and phosphate in the effluent. Frequently, the most convenient method of achieving this is to pass the effluent through a further stage of biological filtration, perhaps using a modified form of percolating filter known as a 'high rate' filter. Nitrogen can be removed by a modified activated sludge process, and ion-exchange processes are available for the removal of phosphate and other inorganic ions, though they are expensive. Sand filters, similar to those used in water supply treatment plants, are effective in improving the quality of the final effluent. Sand filters, of course, are not simply physical filters; as in the percolating filter, the sand particles develop a microbial flora whose metabolism contributes to the effectiveness of the process. An additional stage of simple sedimentation in an appropriate tank has frequently been found to yield a substantial reduction in BOD and suspended solids in the final effluent. Lagoons, similar to oxidation ponds, are a useful form of tertiary treatment where sufficient land is available, and effectively reduce BOD, suspended solids and nutrients. An alternative which is widely employed is simply to allow the effluent to pass slowly over an area of slightly-sloping grassland. Organic solids are deposited in the soil, reducing suspended solids and BOD, and plant nutrients can be effectively absorbed by the growth of grass, which can be periodically mown and removed. The optimum solution for any particular tertiary treatment problem depends primarily upon local conditions.

6.1.4 Sludge disposal

Sewage treatment generates substantial volumes of organically-rich sludge. In stabilisation ponds the sludge itself plays an important part in the treatment process, but in other processes the sludge which is generated following primary and secondary treatment must be disposed of. Various means are employed to treat and dispose of sewage sludge, the method chosen depending mainly upon local conditions and the particular characteristics of the sludge. In general, sewage sludge is composed approximately of 95% water and 5% organic and inorganic particulate matter. Secondary sludge tends to have a higher organic content than primary sludge, and all sludges contain some quantity of potentially toxic material, and various pathogenic organisms.

One approach to the problem of sludge disposal is to regard it simply as waste material. In coastal areas, particularly in Britain and the USA, sewage sludge is frequently dumped at sea from specially-constructed vessels. The extent to which this practice represents a serious threat of marine pollution is a matter of controversy (see Chapter 7). At inland sites, disposal at sea is not practicable and other means of disposal must be sought. A major difficulty is to reduce the bulk of the sludge, and one of the simplest means of achieving this is to spread the liquid sludge onto drying beds, allowing the water to evaporate and consolidating the sludge into a more compact form, so that it can be accumulated on-site or perhaps transported elsewhere for disposal as landfill, as fertiliser or for other useful purposes. In some cases, liquid sludge is transported from the treatment site and spread onto agricultural land, where it may perform a useful function both as fertilizer and as a means of irrigation. Because sewage sludge is, potentially, a valuable resource rather than simply a waste material, many methods have been devised specifically for the treatment of sludge, and in particular for reducing its bulk (for ease of transport), and for eliminating pathogenic organisms. A concise description of sludge treatment processes is given by White (1978); some important biological aspects of sludge treatment are discussed by Mosey (1983) and Green (1983). The methods of sludge treatment applied in particular circumstances depend largely upon local conditions and upon economic considerations.

Sludge-drying beds are formed by spreading a layer of sludge, approximately 200-mm thick, over a drainage layer of ash or similar material. Liquid effluent from the drying beds can be collected by drainage channels and is ideally recirculated for further secondary treatment, as it has a high BOD and organic content. The remaining water is lost by evaporation. Sludge-drying beds require extensive land areas, especially in temperate or cool climates, and can create local nuisances due to odours and insects. Thus there are many advantages in reducing the water content of sludge by other means, and the addition of chemical coagulating agents (such as compounds of iron and aluminium) to sludge has been found useful. Heat treatment of sludge (at about 200°C for 0.5 h) also appears to aid coagulation, and is useful in reducing the load of pathogenic organisms. Various processes of sludge dewatering which rely on filtration under pressure are also widely used. Sludge treatment processes which are essentially biological in nature include anaerobic fermentation and composting. Anaerobic fermentation of sludge (sludge digestion) takes place in closed vessels, and results in a marked reduction in the organic content of the sludge as a result of anaerobic bacterial action. The operating conditions of sludge digesters vary (some, for example, are heated to promote the growth of particular bacteria, or to speed up the digestion process). In many cases, the end product of bacterial metabolism is methane, which can be collected and burned to provide energy for heating the digesters, or for other stages of the sewage treatment process. Composting of sludge is an aerobic process similar to the traditional garden compost heap; typically, sewage sludge is mixed with solid domestic waste in a ventilated enclosure. The end product of the process is an organic humus of value as an agricultural fertiliser. In many conurbations, the disposal of domestic solid wastes and of sewage sludge each pose a problem. Domestic waste has a high organic content but is too dry and frequently too deficient in nitrogen to allow composting to occur. Sewage sludge has too high a water content to allow aerobic decomposition to occur readily. Mixing the two forms of waste together under appropriate conditions can offer an economic solution to both problems. Finally, if no other method is appropriate to local conditions, sewage sludge can, like solid domestic waste, be incinerated.

6.1.5   Biological treatment of industrial wastewater

Although conventional biological treatment processes were devised primarily for the treatment of sewage, they often work well with industrial wastes, though frequently the treatment processes have to be modified. Many industrial wastewaters are similar in composition to raw sewage, particularly those from the dairy and food processing industries, and from industries involved in the processing of natural materials. However, a common problem is that industrial wastewaters are much more concentrated than typical sewage. Dairy wastes, for example, are difficult to treat by conventional biological filtration because the filters readily clog with fatty material and fungal masses. The concentrated nature of dairy effluents means that they rapidly begin to decompose anaerobically, so clogging of the filters rapidly leads to the breakdown of the whole treatment process. A typical means of overcoming this difficulty is to modify the conventional treatment process. Firstly, the raw effluent is diluted; this can be achieved by mixing the raw effluent with a quantity of the final, treated effluent. Secondary treatment is carried out not with a conventional percolating filter, but by a modified process known as alternating double filtration. In this system, two percolating filters are arranged in series. The effluent from the first is passed, after settling, through the second. The biological film in the first filter builds up rapidly; however, before the filter becomes clogged the flow of effluent is diverted so that the second filter receives the raw effluent first, while the first filter now receives partially-treated effluent. This has the effect of stripping out from the first filter the excessive quantity of film, which settles out in the sedimentation tank. When the second filter approaches the stage at which it begins to clog, the filters are alternated again. The regular alternation of the filters brings about a very effective treatment of the highly-concentrated waste; indeed, many conventional sewage treatment works operate on the alternating double filtration system simply to increase the speed of treatment.

A wide variety of industrial wastewaters can be successfully treated by modified biological processes. A common practice is continuously to seed the percolating filter or activated sludge plant with a range of micro-organisms which have been specifically isolated and cultured for their ability to metabolise the waste in question. Frequently, industrial wastes need special pre-treatment before they are amenable to biological treatment. Common forms of pre-treatment include neutralisation of extreme pH; the removal of toxic substances by ion-exchange, coagulation or volatilisation; and the physical separation of oils, greases or other undesirable matter which may interfere with the normal treatment processes. A common difficulty is that many wastes, including some which are principally organic in origin, are deficient in one or more of the nutrients which are essential for the rapid development of the fauna and flora of the treatment process. Frequently nitrogen, phosphorus or other nutrients in a biologically-available form are added to the wastewater to enhance the efficiency of the treatment process.

6.2 LEGISLATIVE AND ADMINISTRATIVE MEASURES

Important as it is, the control of water pollution is only one aspect of the overall management and utilisation of water resources. Generally, the policies and practices associated with water pollution control are broadly compatible with those of water supply, fisheries, conservation and recreation. They may conflict, however, with political or economic priorities associated with waste disposal, industrial or agricultural development, transportation or other legitimate demands on water resources. Pollution control policies must also be formulated in accordance with the available technical, human and economic resources. Therefore the legislative and administrative measures which are employed in water pollution control vary widely from one country to another, and indeed change over time within a single country as circumstances change. This discussion introduces some examples of legislative and administrative approaches to water pollution control. It is based largely upon the procedures adopted in Britain; the British system is not necessarily the best approach to the problem, but it is a well-developed system which has evolved in a relatively wealthy, industrialised and densely-populated country and therefore provides a useful example.

6.2.1   Water management in Britain

Until recently, responsibility for the various aspects of water management in Britain was shared between a large number of different organisations. Broadly speaking, until 1974 water supply was the function of a large number of separate bodies, of which some were municipal undertakings and some private companies. Sewage disposal was the responsibility of municipal authorities around the country; and river management (drainage, flood prevention, fisheries, conservation and pollution monitoring) was undertaken by a number of River Authorities. Separate management of these three basic functions worked reasonably well (although many municipal authorities neglected their sewage treatment works) and is still practised in many countries. However, the potential advantages of a unified management system for all aspects of the hydrological cycle were widely recognised. The Water Act of 1973 led to the establishment of ten Regional Water Authorities in England and Wales in 1974 (Scotland retained the old system), each regional authority being responsible for all aspects of water management within their areas. These areas were defined not by political boundaries, but by the major river catchment areas. These large, unified authorities with their considerable physical, human and financial resources were intended to be better able to cope with the increasingly complex management and technical problems associated with increased demands for water. Arguably, it is easier to devise and implement policies on the interrelated problems of water supply, pollution control and monitoring, conservation, river management, recreation, fisheries and so on within a single organisation. However, one paradox of the unified management system is that the organisation responsible for monitoring and regulating pollutant discharges to water is itself the most important single polluter of water, since the water authorities are themselves responsible for sewage disposal. This difficulty is resolved by requiring that the water authorities, which are normally empowered to grant or refuse permission to discharge wastes to water, require the approval of the Secretary of State (Environment Minister) for their own discharges. Under new legislation introduced to Parliament in November 1988, a new National Rivers Authority is to be established to supervise and monitor the performance of the water authorities.

The powers and responsibilities of the Water Authorities in Britain are defined in a number of laws, including the Rivers (Prevention of Pollution) Acts of 1951 and 1961; the Water Acts of 1945, 1973, and 1983; the Salmon and Freshwater Fisheries Act of 1975; and the Control of Pollution Act of 1974. Generally, the discharge of wastes into water bodies is only permitted with the consent of the water authority. Consent is normally given only if certain conditions are met, such conditions relating to both the quantity and quality of the effluent being discharged, and frequently requiring in addition that the quality, quantity and effects of the effluent be systematically monitored and recorded. Failure to observe the consent conditions can lead to prosecution of the offending discharger. In addition, the water authorities have a general duty to ensure that the waters for which they are responsible do not deteriorate, and indeed that under appropriate circumstances the quality of the waters is actually improved. For this reason, water authorities routinely undertake large-scale programmes of biological monitoring. This activity not only ensures that their statutory responsibilities are being met, but also provides useful data for the formulation of water quality objectives (see section 6.3.4).

6.3   THE REGULATION OF POLLUTION

The legislative framework within which regulatory agencies operate may be broadly similar in many different countries, but in practice the important consideration is the basis upon which consent conditions are formulated. It is on this issue that practices differ widely between different countries, depending upon local circumstances; and it is upon this issue that purely scientific or technical judgements need to be balanced against socio-economic criteria. As pollution is increasingly recognised as an international rather than a purely local problem, such differences in approach become correspondingly more important. To take a simple example, many European rivers pass through several different countries and differences in pollution control strategies can give rise to disagreement between sovereign states. In Europe, member states of the European Economic Community (EEC) are subject to EEC laws and regulations which may dictate departures from their normal pollution control practices; differences which may arise between politicians from the different countries concerned are no more rigorously debated than those which occur between the scientists! Similar situations arise in the USA and in other federal countries, where state regulatory agencies may be subject to federal laws or regulations which are different in approach or emphasis from state laws or practices. It is therefore important to consider the various approaches which may be taken to the problem of what constitutes an 'acceptable' level of pollution.

For the purposes of this discussion, four principles can be distinguished as forming the basic elements of an effective strategy of pollution control. These are:

(a) The 'polluter pays' principle.
(b) Receiving water quality standards.
(c) Emission standards.
(d) Water quality objectives.

6.3.1   The 'polluter pays' principle

It is widely accepted that those who are responsible for pollution should contribute towards the costs of monitoring, regulating and controlling the adverse effects of pollution. However, the 'polluter pays' principle is clearly insufficient in itself as an effective and equitable mechanism of environmental protection. Consider, for example, a factory which produces a substantial amount of contaminated wastewater. The factory owners may be unwilling to meet the costs of treating the wastes, or may lack the technical ability to do so. The factory therefore discharges untreated waste, thereby obtaining an economic benefit at the expense of the community at large. Clearly the factory should be made to compensate the community for this, perhaps passing on the economic charge to the eventual consumers of its products. The calculation of the charge to be made, however, is difficult. It should be possible to calculate the economic cost of treating the waste, and the extra costs incurred at water treatment works owing to the fact that the raw water is of lesser quality than it would otherwise have been. The additional costs to the regulatory authorities of monitoring the effects of the resulting pollution, and the economic losses associated with any deterioration in fisheries, are in principle calculable, though not easily so. However, it is difficult to estimate in purely economic terms the amenity or aesthetic value of unpolluted water in comparison with polluted water, or to put a cost in money terms on the loss of species diversity in the aquatic community. It must also be recognised that not all polluters can be identified and charged. Further, there are several objections to the idea that permission to pollute should be given in exchange for a cash contribution to the community at large. In the absence of some additional form of regulation, the 'polluter pays' principle leads almost inevitably to the idea that pollution is a right which may be exercised by anyone who is willing and able to pay the appropriate charge. There are moral and philosophical arguments against such a position, but in any case the idea is impracticable. For example, some rivers may be already so badly polluted that any additional burden would create ecological or public health consequences which are unacceptable at any price. Some forms of pollution, such as radioactivity and certain toxic chemicals, are so potentially hazardous that their disposal demands special procedures; to allow their discharge simply in exchange for an economic charge would be extremely dangerous. Finally, some habitats are of particular scientific, conservation or amenity value and their protection from pollution would be undermined by acceptance of the simple proposition that pollution can be compensated for by a cash payment. Thus while considerations of logic and equity dictate that elements of the 'polluter pays' principle should be preserved, the principle must operate within a regulatory framework which prevents the inherent tendency of the principle to lead to environmentally undesirable consequences.

Some examples from British regulatory practice illustrate one way in which the 'polluter pays' principle can be incorporated into an overall pollution control strategy. For example, every household pays a charge for the water it consumes which is distinct from general taxation, and includes also a charge for the treatment and disposal of sewage and for environmental services (i.e. pollution monitoring, fisheries, river management and similar activities). The discharge of wastes to water bodies is expressly forbidden except with the permission of the water authority. Permission can be refused altogether, to prevent the discharge of particular hazardous wastes or to protect specific habitats from pollution. Where permission is granted, qualitative and quantitative limits are set which must not be exceeded. If the limits are exceeded, the 'polluter pays' principle comes into operation in the form of fines and penalties imposed by the courts on the offender. These economic penalties are often in themselves relatively insignificant, but offenders can also be compelled to pay for the restoration of any damage they cause (such as restocking rivers with fish), and this cost is often the most signifiant economic penalty. Water supplies to industrial premises are charged for at a rate which may reflect the costs of treating the wastewater, of monitoring the impact of the discharge on the receiving water, of increased costs of managing and maintaining the receiving water and of increased water treatment costs of downstream water supply works. Large undertakings frequently undertake their own waste treatment, either in exchange for reduced water supply costs or because the consent conditions for their discharges make this an absolute requirement. Smaller undertakings are permitted, within limitations, to discharge wastes to the public sewage system in exchange for a payment. The level of these payments can, at least in principle, be set at levels which act as an economic incentive to the factory operators to improve the quality or reduce the quantity of the effluent. Thus the 'polluter pays' principle operates only within regulatory constraints which are intended to ensure that the principle has beneficial rather than adverse environmental effects.

6.3.2　Receiving water quality standards

In order to decide whether or not a polluting discharge should be permitted, and if so under what conditions, it is obviously necessary to have some idea of the maximum quantity of pollutant which a water body can assimilate without undergoing adverse consequences. Where the receiving water is to be used as a source of water supply, or for recreational purposes or agricultural irrigation, clearly special considerations may apply. However, for the purposes of this discussion it will be assumed that the dominant consideration is the preservation of aquatic life. The degree of pollution which is deemed 'acceptable' in any particular circumstances is only partly a biological question, but a reasonable objective which may be adopted is that the existing biological characteristics of the receiving water should not be significantly altered by the discharge of wastes to the receiving water. To implement this water quality objective (see section 6.3.4) by means of specific decisions on the quantity and quality of effluent which may be discharged requires a reasonably reliable estimation of the likely effects of pollutant concentrations on individual species within the aquatic community, on the community as a whole and on the physicochemical environment of the receiving water. If it can be established, for example, that a certain concentration of a toxic substance is unlikely to exert any significant harmful effect in the receiving environment or upon its biota, then discharges can be regulated by various means to ensure that the concentration of the poison in the receiving environment is kept below that level.

Water quality standards for different forms of pollution are devised in different ways, because the means by which pollutants bring about their effects vary. Putrescible organic matter, for example, brings about a variety of changes in the receiving water environment (see Chapter 2) but these can largely be avoided if the dissolved oxygen concentration is maintained above a certain level. Knowing the BOD of the effluent, the degree of dilution afforded by the receiving water and certain physical characteristics of the receiving environment, it is possible to predict fairly accurately the oxygen sag curve which will occur. The dissolved oxygen requirements of many important aquatic species, especially fishes, are reasonably well known (see for example, Alabaster & Lloyd, 1980). Therefore it is relatively simple to determine a quality standard for dissolved oxygen which can be translated into an emission standard (see section 6.3.3) for organic matter. Toxic pollutants present greater difficulties, mainly because relatively little information exists on the effects of long-term, low level exposure of animals to many poisons. Additionally, the toxicity of poisons is greatly influenced by environmental conditions; many poisons are selective in their effects; and they normally occur in water in mixtures and at fluctuating concentrations (see Chapter 4). Water quality standards for toxic pollutants are derived on the basis of evidence from experimental studies and from observations made during chemical and biological studies of polluted waters. For all but the most common pollutants, however, relatively little data is available from field studies and water quality standards have to be based on toxicological data. This is obviously also true in the case of novel pollutants.

The bulk of the available toxicological data relates to lethal levels of pollutants; much less information is available on sublethal toxicity and the behaviour of pollutants at realistic levels in, for example, controlled or semi-controlled ecosystems. Data of the latter kind are accumulating slowly, as techniques are developed, but the main problem of formulating water quality standards from toxicological data remains that of using median lethal concentrations as the basis for estimating concentrations which will not only fail to kill any organisms, but which will also allow them to survive, grow and reproduce normally. One widely-used technique for deriving water quality standards from toxicity data is the 'application factor'. It is relatively simple and quick to determine lethal toxicity. The application factor is based on the assumption that for every poison there must be a concentration which is so low that it has a negligible effect; and that this concentration should bear some approximate relation to the lethal threshold concentration (see Chapter 4). Thus, multiplying the lethal threshold concentration by a factor (e.g. 0.1) should give a rough indication of the 'acceptable' level of the poison.

Methods for deriving application factors vary, and have been briefly reviewed by Sprague (1971). Some methods are little more than simple extrapolation of the data, others are based on statistical or other considerations, but all are subject to a degree of arbitrariness. There is little scientific validity in their use, but in practice they have been found useful and many regulatory agencies in Europe and North America have published recommended application factors for a wide range of poisons. Values vary from about 0.01 for persistent or cumulative poisons, to about 0.3 for rapidly-degradable pollutants of fairly low toxicity. As Sprague (1971) emphasised 'The value of the application factor is assigned on the basis of the judgement of scientists.' Over the last 15 years or so, the scope and quality of the available data has improved so that although scientists still use their judgement, they appear to have less need of the concept of the application factor, and the more recent literature contains few references to the idea. This seems to have occurred as a consequence of the development of sublethal toxicity testing techniques, of controlled ecosystem studies and of improved data from field studies. Early exponents of sublethal toxicity testing frequently used their data to determine empirically the values of application factors, for example by calculating the ratio of the 'No Observed Effect Concentration' (NOEC) to the lethal concentration. However the determination of NOEC is itself a somewhat arbitrary process (see Chapter 4); and the application factor approach has more recently been superseded by what has come to be known as sequential hazard assessment (Duthie, 1977).

Sequential hazard assessment is not in fact a particularly novel approach to assessing the potential effects of water pollutants. Rather, it is simply a codification of the sequence of investigative processes which has long been recognised as necessary to derive water quality standards from toxicological data. The development of modern techniques, and very importantly the gradual accumulation of a reliable data base of previous experience, render sequential hazard assessment possible; but the underlying process of scientific judgement remains the same as it always has been. The approach has been described, with examples, very clearly by Tooby (1978), and can be expressed in the form of a simple flow chart (Fig. 6.5).

Essentially, a sequential hazard assessment scheme consists of a number of steps, each providing progressively more complex information. Each succeeding stage in the investigation is generally more expensive and time-consuming than the previous one. The advantage of such a scheme is that at least in some cases, the later and more difficult steps can be dispensed with. For example, Fig. 6.5 shows that the preliminary assessment is designed to determine the 'margin of safety'. Here, the results of a few simple screening tests can be compared with the projected environmental concentrations and patterns of use in the chemical. If, say, the 96 h LC50 of the chemical to a carefully-chosen range of test species is more than 10000 times greater than the expected environmental concentration, experience suggests that the chemical is not particularly hazardous. Taking into account its pattern of use (e.g. whether it will be permanently present in the water, or only for a few hours or days occasionally), the decision may be made that no further tests are required and that permission to discharge it may be given, at least provisionally. If, on the other hand, the 96 h LC50 values are within an order of magnitude of the expected environmental concentration, the likely decision will be that the chemical is unacceptable, again without the need for further tests. Margins of safety lying between these two extremes would tend to indicate the need for further investigations, that is to proceed to the next step of the scheme.

At any stage, the investigation can be terminated when sufficient information has been gained to make, in the light of previous experience, a decision. Both industry and regulatory authorities can follow a similar procedure, and the advantages in terms of speed and economy are obvious. The procedure not only allows the simple decision 'accept or reject the chemical' to be made. It clearly also permits decisions to be made upon, for example, the concentrations which are acceptable and any restrictions or conditions which should be put on the discharge of the pollutant (e.g. to protect early life stages at certain times of year). Obviously, however, any water quality standard determined on this basis is tentative, and should ideally be evaluated in the light of ecological monitoring of the receiving waters subsequent to the discharge of the pollutant.

In the case of common industrial pollutants of long standing, where for example it is desired to improve the status of a polluted receiving water, the approach to determining water quality standards is often significantly different. A large amount of toxicological information may already exist, and data from field observations are likely to assume greater significance. A very good illustration of the considerations which may be taken into account in formulating water quality standards is the series of reviews edited by Alabaster & Lloyd (1980). These reviews recommend water quality standards for the protection of European freshwater fisheries against several common industrial pollutants. It is also useful to recall the very sophisticated approach to formulating water quality standards described in section 1.2. The common industrial pollutants copper, zinc, phenol, ammonia and cyanide commonly occur together, and their toxicity is greatly influenced by environmental conditions.

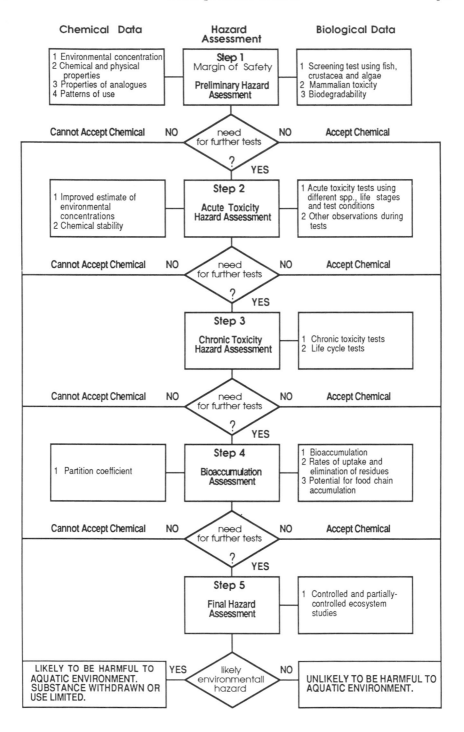

Fig. 6.5 — Flow chart summarising the steps in the sequential hazard assessment procedure. After Tooby (1978).

The approach described in section 1.2 offers the means to identify specific threats to, or potential improvements in, the biological status of a receiving water which can be fairly easily implemented by specific regulatory action against polluting discharges or by alternative strategies of river management. The successful application of this method requires, however, an extensive data base of chemical, biological and toxicological information.

6.3.3 Emission standards

Having determined the appropriate receiving water quality standards, it is next necessary to ensure that pollutant concentrations in the receiving water do not exceed the critical levels. Taking into account the volume of water available to dilute the effluent, and the quantity and strength of the effluent, it is possible to place qualitative and quantitative limits on the amount of effluent which may be discharged. An early example of an emission standard in Britain was the recommendation of the Royal Commission on Sewage Disposal in 1915 that sewage effluent should contain no more than 30 mg l^{-1} suspended solids and should have a BOD value not exceeding 20 mg l^{-1}. This recommendation (the so-called 30/20 standard) was based on the assumption that the effluent would normally be diluted at least eight times by the volume of the receiving water. In practice, consent conditions for the discharge of effluents may set limits on the strength and quantity of effluent which may be discharged; individual components of the effluent may be subject to specific restrictions; and consent conditions may vary according to the time of year, or the physical conditions of the receiving water such as temperature or the volume of dilution water available.

It is comparatively easy to monitor compliance with emission standards, and they have often been found useful even when set at more or less arbitrary levels. For example, the 30/20 standard resulted in a substantial improvement in the quality of many British rivers, even if the standard was not universally achieved, because at the time it was implemented most sewage was discharged either untreated, or treated only to a poor standard. Emission standards for oils and toxic substances can usefully be applied to effluents in which they are only rarely to be expected (for example as a result of plant failure), since the existence of the standard will require the effluent to be monitored and will thus tend to prevent pollution caused by unusual occurrences which may otherwise be undetected. Emission standards are also administratively convenient. Consider, for example, a river which is already polluted to which it is proposed to discharge a new effluent. The regulatory agency may decide that the river cannot accommodate a substantial increment of pollution, and may wish to set consent conditions for the new factory which are more restrictive than those which apply to the established discharges. This can lead to litigation or political lobbying from the owners of the new factory, who may feel that they are being treated inequitably in comparison with their competitors. An equitable and administratively convenient solution would be to revise the consent conditions of all the factories concerned so that each factory was subject to the same consent conditions — in other words, uniform emission standards would be imposed on all.

The administrative convenience of uniform emission standards tends to lead to their assuming a degree of importance in pollution control strategies which can be counter-productive. This problem is becoming more acute as pollution control strategies become internationalised. Within the European Community, for example, member states have agreed in principle to adopt policies designed to reduce environmental pollution. Different states within the Community may each feel that their particular pollution control strategy is adequate in relation to local conditions. Inevitably, however, emission standards will vary from one place to another, and some states will be able to argue that the stricter constraints under which they operate impose upon them an unfair economic disadvantage. It is therefore difficult, politically and administratively, to avoid moving towards a policy of uniform emission standards regardless of local conditions. A similar situation can arise where several states enter into agreements to limit pollution in rivers, lakes or seas to which all discharge effluents.

The principal disadvantages of uniform emission standards are that they are not always appropriate for the protection of the receiving environment, and that they can lead to the inefficient use of resources. For example, a sewage works discharging to a river which is already heavily polluted could operate to a comparatively low standard of effluent quality and still actually improve the quality of the receiving water, perhaps leaving resources to spare which could be better utilised elsewhere. Conversely, a new discharge which meets a uniform emission standard could have a seriously adverse effect if sited on a river which was previously unpolluted. For this reason, in Britain and several other countries a fourth element of pollution control strategy has been developed, the concept of water quality objectives.

6.3.4 Water quality objectives

The concept of pollution regulation by water quality objectives (WQOs) is based on the argument that the most effective use of resources and the greatest degree of protection and improvement of receiving waters is in practice achieved by the formulation of emission standards which are designed to be appropriate to the particular receiving water under consideration. In other words, uniform emission standards are specifically rejected. Instead, each receiving water is considered individually, and a technically and economically feasible management objective is determined. An example of such an objective may be that the receiving water should sustain a healthy population of cyprinid fish. The appropriate receiving water quality standards are then determined, and can be translated into emission standards designed to achieve the initial objective.

The principle difficulty of the WQO approach is that it can only be sensibly and effectively applied on the basis of extensive and reliable information about the current status of receiving waters and of the pollution loads which they are receiving. Without this information, feasible water quality objectives cannot be set, and if the objective is not realistic the resulting emission standards will be as arbitrary and inefficient as uniform emission standards. For example, it may be highly desirable that all rivers sustain populations of salmonid fishes and a diverse invertebrate fauna. Such an objective ignores the fact that many rivers are, for various reasons, naturally unable to do so. Also, regulatory agencies are more likely to enjoy the co-operation of industry and the support of the general public if the economic resources devoted to pollution control result fairly quickly in a perceptible improvement in environmental quality. It is therefore important that water quality objectives are realistic in relation to the current status of the receiving waters.

In Britain, the application of WQOs to pollution control rests upon extensive national surveys of all substantial rivers. These surveys began about thirty years ago, and now take place every five years. The survey work is carried out by the Regional Water Authorities and co-ordinated by the Department of the Environment. An important aspect of the national river surveys is the development of a national scheme for the collection and analysis of data. This was a major factor in the development of the BMWP Score, a biotic index of water quality devised to be appropriate in all parts of the country and in all types of river (see section 3.4.3). On the basis of the survey information, a national system of river classification was devised (National Water Council, 1981) which can be used to formulate sensible water quality objectives. The classification scheme is shown in Table 6.2. The classification immediately suggests several objectives. Class X rivers, for example, are defined as insignificant watercourses of no potential use other than for drainage. Here, a rational objective is simply to prevent them becoming a health hazard or a public nuisance. For a class 1A river, however, an obvious objective is to prevent any deterioration at all. For rivers of intermediate classification, a reasonable objective is at least to preserve them in their present condition, and where feasible to improve them so that they attain the next highest classification. In a typical scheme devised by the Severn-Trent Water Authority (Young, 1980) it was proposed to leave class X rivers as they are; to eliminate class 4 rivers by improving them to class 3 quality; to improve selected rivers of classes 3 and 2 to classes 2 and 1B respectively; and to preserve class 1A rivers in their present condition (Table 6.3).

Note that an important aspect of the classification system is the current or potential uses of the water, which both determine the present classification of the water and help to identify those rivers for which improvements would be most cost-effective. Prior to 1980, a different classification scheme was used, based largely upon physical and chemical criteria. The classes of the old scheme are referred to in Table 6.3 as RPS (River Pollution Survey) classes. The new classification scheme differs essentially in that the basis of the classification gives rather more weight to the actual and potential uses of the water, and to biological characteristics; and the old Class 1 has been split into Classes 1A and 1B, while the old Class 4 has been replaced by Classes 4 and X. Table 6.4 summarises the classifications of non-tidal rivers and canals in England and Wales in national surveys between 1958 and 1980, clearly demonstrating a steady overall improvement in water quality during that period. The report of the 1985 survey (Department of the Environment, 1986) shows that the rate of improvement has slowed, and that some rivers have clearly deteriorated. However, the report argues that this is at least partly due to changes in survey methodology introduced in 1980.

Pollution control by quality objectives is arguably the most scientifically and economically sophisticated approach to the problem of devising an effective system for regulating waste discharges. However, its effectiveness and public acceptance depends crucially upon extensive and accurate information. This in turn depends upon the establishment of nationally-based or regionally-based systems of monitoring and co-ordination, and of an appropriate technical and administrative infrastructure; these are, unfortunately, well-established in relatively few parts of the world. It is for this reason that regulatory strategies differ from one country to another, although these differences primarily consist of differences in the relative emphasis

Table 6.2 — The National Water Quality Classification Scheme used in Britain

River class	Quality criteria	Remarks	Current potential uses
1A Good quality	Class limiting criteria (95 percentile) (i) Dissolved oxygen saturation greater than 80% (ii) Biochemical oxygen demand not greater than 3 mg/l (iii) Ammonia not greater than 0.4 mg/l (iv) Where the water is abstracted for drinking water, it complies with requirements for A2[a] water (v) Non-toxic to fish in EIFAC terms (or best estimates if EIFAC figures not available)	(i) Average BOD probably not greater than 1.5 mg/l (ii) Visible evidence of pollution should be absent	(i) Water of high quality suitable for potable supply abstractions and for all other abstractions (ii) Game or other high class fisheries (iii) High amenity value
1B Good quality	(i) DO greater than 60% saturation (ii) BOD not greater than 5 mg/l (iii) Ammonia not greater than 0.9 mg/l (iv) Where water is abstracted for drinking water, it complies with the requirements for A2[a] water (v) Non-toxic to fish in EIFAC terms (or best estimates if EIFAC figures not available)	(i) Average BOD probably not greater than 2 mg/l (ii) Average ammonia probably not greater than 0.5 mg/l (iii) Visible evidence of pollution should be absent (iv) Waters of high quality which cannot be placed in Class 1A because of the high proportion of high quality effluent present or because of the effect of physical factors such as canalisation, low gradient or eutrophication (v) Class 1A and Class 1B together are essentially the Class 1 of the River Pollution Survey (RPS)	Water of less high quality than Class 1A but usable for substantially the same purposes

Class			
2 Fair quality	(i) DO greater than 60% saturation (ii) BOD not greater than 9 mg/l (iii) Where water is abstracted for drinking water it complies with the requirements for A3ᵃ water (iv) Non-toxic to fish in EIFAC terms (or best estimates if EIFAC figures not available)	(i) Average BOD probably not greater than 5 mg/l (ii) Similar to Class 2 of RPS (iii) Water not showing physical signs of pollution other than humic colouration and a little foaming below weirs	(i) Waters suitable for potable supply after advanced treatment (ii) Supporting reasonably good coarse fisheries (iii) Moderate amenity value
3 Poor quality	(i) DO greater than 10% saturation (ii) Not likely to be anaerobic (iii) BOD not greater than 17 mg/l. This may not apply if there is a high degree of re-aeration	Similar to Class 3 of RPS	Waters which are polluted to an extent that fish are absent or only sporadically present. May be used for low grade industrial abstraction purposes. Considerable potential for further use if cleaned up
4 Bad quality	Waters which are inferior to Class 3 in terms of dissolved oxygen and likely to be anaerobic at times	Similar to Class 4 of RPS	Waters which are grossly polluted and are likely to cause nuisance
X	DO greater than 10% saturation		Insignificant watercourses and ditches not usable, where the objective is simply to prevent nuisance developing

Notes (a) Under extreme weather conditions (e.g. flood, drought, freeze-up), or when dominated by plant growth, or by aquatic plant decay, rivers usually in Class 1, 2 and 3 may have BODs and dissolved oxygen levels, or ammonia content outside the stated levels for those Classes. When this occurs the cause should be stated along with analytical results.
(b) The BOD determinations refer to 5-day carbonaceous BOD (ATU). Ammonia figures are expressed as NH_4.
(c) In most instances the chemical classification given above will be suitable. However, the basis of the classification is restricted to a finite number of chemical determinands and there may be few cases where the presence of a chemical substance other than those used in the classification markedly reduces the quality of the water. In such cases, the quality classification of the water should be down-graded on the basis of biota actually present, and the reasons stated.
(d) EIFAC (European Inland Fisheries Advisory Commission) limits should be expressed as 95 percentile limits.

ᵃ EEC category A2 and A3 requirements are those specific in the EEC Council directive of 16 June 1975 concerning the Quality of Surface Water Intended for Abstraction of Drinking Water in the Member State.

Table 6.3 — Proposed river quality objectives for the Severn-Trent Water Authority area in 1978. (Data from Young, 1980)

Classification	Existing quality km	Proposed quality km
1A	877 (14.3%)	877 (14.3%)
1B	2417 (39.5%)	2543 (41.5%)
2	2085 (34.1%)	2360 (38.6%)
3	565 (9.2%)	292 (4.8%)
4	128 (2.1%)	0 (0%)
X	46 (0.8%)	46 (0.8%)

Table 6.4 — Summary of the results of the British National River Surveys, 1958–1980. The table shows the number of kilometres of river in each of the four classes of the pre-1980 classification scheme in each of the six surveys carried out during the 1958–1980 period. Data from National Water Council (1981)

Class	1958	1970	1971	1972	1975	1980
1	24950	28500	28960	29280	28810	28810
2	5220	6270	6250	5960	6730	7110
3	2270	1940	1750	1770	1770	2000
4	2250	1700	1470	1400	1270	810
Total	34690	38410	38430	38410	38580	38730

given to the 'polluter pays' principle, water quality standards, emission standards, and water quality objectives. Some of the issues involved are discussed in the volumes edited by Stiff (1980) and Lack (1984).

7

Estuarine and marine pollution

The effects of water pollution, and their consequences for public health, became obvious in lakes and rivers long before any general threat to estuaries and seas was widely perceived. Consequently the methodology and literature relating to the study of estuarine and marine pollution are less well-developed, and many of the basic approaches to the study of pollution in maritime habitats are derived from ideas developed for fresh-water habitats. Certainly, the answers sought from the study of fresh water, marine and estuarine habitats are essentially the same; nevertheless there are significant physical, chemical and biological differences between fresh water and maritime environments. Because of these differences, the technical and conceptual approach to the study of pollution in maritime habitats must be different, at least in emphasis, to those which are appropriate to the study of fresh waters. This chapter introduces some of the special problems of pollution which occur in estuaries and in the sea, and considers the extent to which the concepts and methodology which are appropriate for fresh waters can be applied to the study of pollution in maritime habitats. A good general introduction to the problems of marine pollution is given by Clark (1986), and more detailed reviews of some important aspects may be found in Kinne (1984a,b).

During the present century social, economic and industrial trends have combined to increase the threat of pollution to estuaries and coastal seas. Most of the world's major cities are now situated on estuaries or close to the coast. Increasingly, large industrial sites are required to be placed away from centres of population, but close to sources of water for cooling and waste disposal, and with access to transportation facilities (e.g. by ships of steadily-increasing size) for the import and export of raw materials and of finished products. Thus major industrial development increasingly takes place in the lower reaches of estuaries or at coastal sites, often on reclaimed land. In many parts of the world, the availability of relatively cheap international travel has led to the development of an economically-important tourist industry. This has increased the pressures of urbanisation upon coastal areas especially; but at the same time it has created strong economic, aesthetic and public health pressures in favour of uncontaminated beaches and coastal seas. Additionally, it has become necessary to consider whether the seas themselves, in spite of their vastness, can accommodate the wastes of which they are the ultimate recipient.

7.1 POLLUTION OF ESTUARIES

Estuaries vary so widely in their characteristics that it is difficult to define them satisfactorily, but Barnes' (1984) definition is a very useful one: 'An estuary is a region containing a volume of water of mixed origin derived partly from a discharging river system and partly from the adjacent sea; the region usually being partially enclosed by a land mass'. In order to understand the effects of pollution in estuaries, it is first necessary to consider some of their special characteristics. Concise descriptions which emphasise those characteristics of estuaries which are particularly relevant to their response to pollution include those of Barnes & Green (1972), Perkins (1974), Arthur (1975) and Barnes (1984).

Barnes (1984) recognises four main types of estuary, based on their geomorphological characters. These are:

(a) Drowned river valleys (coastal plain estuaries), formed by the rise in sea level which occurred at the end of the last glaciation, about 10 000 years ago.
(b) Tectonically-produced estuaries, formed by the subsidence of land and consequent invasion by the sea.
(c) Fjords, glacially-overdeepened valleys into which the sea penetrates. These frequently have a sill at their mouth which greatly restricts the interchange of water between the estuary and the open sea, so that the volume of water lying within the fjord but below the level of the sill is effectively isolated, and behaves rather like a lake.
(d) Bar-built estuaries, formed by the deposition of sand or shingle in a line parallel to the shore which blocks or diverts a discharging river system, forming an estuary in a former area of sea between the land and the sand or shingle bar.

The precise physical characteristics of an estuary greatly influence the biological processes which occur within the estuary, to the extent that it is difficult to generalise about the biological properties of estuaries and the way in which pollution affects them. In some estuaries, for example, the volume of fresh water discharged may greatly exceed the amount of sea water intrusion; in others, the reverse is the case. The ratio of fresh water to sea water, their patterns of mixing and the consequent influences upon the biological characteristics of the estuary depend ultimately upon the shape and size of the estuary, the magnitude of the tidal influences, and the geomorphology and geochemistry of the area. This discussion attempts to describe some relevant aspects of representative coastal plain estuaries (the most common type).

The precise pattern of mixing of fresh water and sea water in estuaries varies greatly, depending upon geomorphological factors. However, all estuaries display both a longitudinal and a vertical gradient of salinity (Fig. 7.1). Further, the salinity of the water at a given point in the estuary tends to vary greatly from hour to hour, according to the state of the tide. Since most living organisms do not tolerate rapid and wide fluctuations in the salinity of their environment, this means that the number

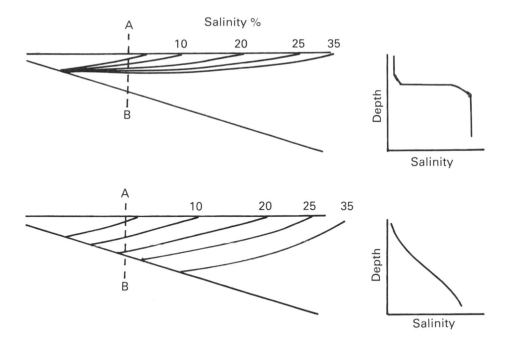

Fig. 7.1 — Horizontal and vertical salinity profiles of two estuaries. In the upper diagram, the position of the isohalines indicates little vertical mixing, the fresh water tending to flow out above a salt-water intrusion (salt-wedge). There is a sharp discontinuity in the vertical salinity profile through the section A–B. The lower diagram represents an estuary where vertical mixing is more pronounced.

of species which can live under estuarine conditions is restricted by comparison with the biota of the open sea, or of fresh water. The majority of species found in estuaries are either fresh water organisms which penetrate the estuary downwards, and which become progressively less numerous as they reach their limits of tolerance; or marine organisms which, in similar fashion, become numerically depleted as they penetrate further upstream. Physicochemical processes within the estuary also produce conditions which limit the biota. By the time a river reaches the sea, most of its load of suspended matter has been deposited, and only the finest particulate matter remains in suspension. The mixing of fresh water with sea water involves a marked increase in the pH and in the level of dissolved salts; these promote the coagulation of fine particulate matter, and of organic and inorganic colloidal matter (Phillips, 1972). Since, during each tidal cycle, the water mass is sometimes moving in one direction and sometimes in the opposite direction, it follows that at every point in the estuary at some time the net water velocity is zero. These processes encourage the deposition of fine organic and inorganic matter on the bed of the estuary. Since most estuaries are also sheltered from wind and wave action, particulate matter brought into the estuary with the incoming sea water also tends to be deposited within the estuary. The bed of an estuary therefore typically consists of fine silt with a high organic content, conditions which many animals will not tolerate, particularly since estuarine muds will tend to be hypoxic owing to the high biological oxygen demand (see section 2.3) imposed by its organic content. Plants, whether macrophytic or planktonic, will also be limited because the high turbidity characteristic of estuarine waters inhibits photosynthesis. In addition, it appears that few animals and plants have evolved specifically to estuarine conditions. One reason for this may be that, geologically speaking, estuaries are short-lived features; fluctuations in sea level and the corresponding changes in the position of the coastline means that an estuary rarely occupies the same position for more than a few tens of thousands of years. The estuarine biota is derived from the fresh water and marine organisms which penetrate the estuary from either end. At the seaward end, the biota is similar to that of the adjacent sea, but species die out rapidly with increasing distance from the sea as they reach their limits of tolerance to decreasing salinity. A similar process occurs to limit the penetration of fresh-water species from the landward end. These factors combine to create the typical situation whereby the estuary has a varied biota at its seaward and landward extremes, but species diversity is characteristically low in the middle reaches. Nevertheless, the handful of species which can survive in the estuary proper are often found in great abundance. The enormous numbers of sea birds which congregate in estuaries at low tide, to feed upon the abundance of invertebrate life, are readily observable. Similarly, estuaries are important sources of food for fish which visit at high tide. The export of biomass from estuaries is almost certainly very important in sustaining populations of ecologically and commercially important species which are not themselves actual residents of the estuary. In addition, many estuaries yield large quantities of shellfish of high commercial value, either from exploitation of natural populations or by means of aquaculture, as well as being crucial in the life cycle of migratory fish.

Given these natural characteristics, it is not difficult to recognise that the principal pollution threat in estuaries is likely to arise from deoxygenation of the water column and sediments due to the accumulation of organic matter, and the accumulation of toxic materials in the sediments. Many polluted estuaries have been studied in detail, and in general this has been found to be the case. In Britain, several of the largest and most industrialised estuaries have been under investigation for about fifty years (e.g. Department of Scientific and Industrial Research, 1935, 1964; Royal Commission on Environmental Pollution, 1972; Porter, 1973). In some of them pollution control measures applied in recent years have led to dramatic improvements, particularly in the Thames. While the situation varies in detail from one estuary to another (and these differences in detail are very significant in the design and implementation of appropriate pollution control strategies), some important general features emerge from these and similar studies. Firstly, it is clear that the long-held assumption that polluting matter is rapidly flushed from an estuary is wrong. Flushing times depend upon the topography of the estuary, the amount of fresh water input and the degree of mixing which occurs between the sea water and the fresh water. However, in most cases flushing times are relatively long, and in some cases essentially the same body of water moves up and down the estuary, according to the tidal cycle, for days or even weeks on end. In addition, indentures and embayments of the estuary shoreline can entrap polluting material more or less permanently. Partial deoxygenation of the water and sediments is a normal phenomenon of many unpolluted estuaries, but in polluted estuaries extensive sections become completely anoxic, either at regular intervals or more or less permanently (Fig. 7.2), and are consequently devoid of higher forms of life. The accumulation of

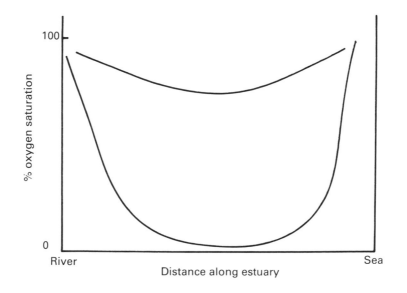

Fig. 7.2 — Oxygen sag curves in (upper line) and unpolluted estuary and (lower line) a heavily-polluted estuary.

toxic matter in the sediments undoubtedly contributes to the elimination of species. Valuable sport and commercial fisheries are eliminated. The predominance of anaerobic micro-organisms, producing hydrogen sulphide and methane as by-products of their respiration, is both inimical to the normal estuarine biota and aesthetically unpleasant. Combined with the obvious presence of tar balls, faecal matter, vegetable and domestic wastes and sewer scum, this effect destroys the recreational and commercial potential of the estuarine shores. The estuary of the river Tees in northern England provides a salutary, but unfortunately by no means unique, example. In the early years of the present century the river supported migratory (salmonid) fish which were exploited commercially and for sport. Above the estuary even today the river is very little polluted, but industrial development of the estuary (Porter, 1973) including substantial petrochemical, agrochemical and steel-making complexes together with dramatic population increases, eliminated the fishery by 1937 (Perkins, 1974). A major survey of the estuary (Department of Scientific and Industrial Research, 1935) recorded the virtual absence of life from the most polluted reach, extending up to 6 km in the central part of the estuary. Severe deoxygenation of the water was a more or less permanent feature, and on occasions the estuary water was rapidly lethal to fish. A more recent survey carried out in 1967 showed that in many respects the position had actually worsened (Porter, 1973).

Fortunately here and in some other British estuaries, remedial action is now being implemented, with considerable success but at great cost.

7.2 POLLUTION OF THE SEAS   (← solo)

Through the direct discharge of wastes to the sea, the discharge of polluted rivers into the sea, from land runoff and atmospheric fallout, the seas are the ultimate recipient of most pollution in most of its various forms. It is now very obvious that extensive areas of sea, particularly in coastal areas, are very seriously affected. What is not clear, however, is the extent to which these are purely local problems which are amenable to local solutions, such as improved waste treatment or the more effective dispersal of wastes over a wider area; or alternatively, the extent to which the severe and obvious local disturbances have more subtle adverse consequences further afield, possibly involving alterations to the entire marine ecosystem. It is virtually impossible to assess the ecological significance of the substantial quantities of some potentially dangerous pollutants which enter the seas through atmospheric fallout. Except in the case of the most severe incidents of pollution, it is very difficult clearly to distinguish the effects of pollution from the natural fluctuations and variations which are known to occur in marine populations and communities. This is true even of communities and species which have been extensively studied on account of their ecological or commercial importance. Nevertheless the quantity and variety of polluting materials which now enter the seas is so great that the long-standing assumption that the seas are sufficiently vast to act as a waste sink of infinite capacity must be questioned. To determine the extent to which the sea's capacity to absorb waste matter is being approached or exceeded is the central question facing biologists in the study of marine pollution. A brief discussion of some of the commoner forms of marine pollution will illustrate the difficulties.

7.2.1 Oil pollution

The public perception of oil pollution as a major threat to the sea has been distorted by the occurrence of a relatively small number of major accidents (shipwrecks, oil-well blowouts), by the enormous tonnage of oil transported around the world, by the vigorous and expensive efforts made to clear oil spills (more often for aesthetic than ecological reasons) from beaches, and by the fact that sea birds, in substantial numbers, are often the most conspicuous casualties of major spills. Detailed review of the scientific evidence, such as those given by Johnston (1984) and in the volume edited by Clark (1982a), afford a more realistic perspective. Firstly, only about one third of all the oil which enters the sea does so as a result of activities associated with oil transportation, and of that only about one quarter is due to accidents and major spillages; the remainder is accounted for by normal operational losses. About 45% of the oil which reaches the sea does so from polluted rivers, urban runoff, municipal wastes and effluents from non-petroleum industries. Up to 20% of the total enters the sea from natural oil seeps, that is locations where oil naturally escapes from under the sea bed. Finally, oil is produced, ultimately, by living organisms, and hydro-carbon production by contemporary populations of marine phytoplankton almost certainly far outweighs anthropogenic inputs of hydrocarbons to the sea. On the basis of figures cited by Davenport (1982), hydrocarbon production by phytoplankton is at least thirty times greater than the anthropogenic input. Johnston (1984) estimates that five million tons of oil enter the sea annually from anthropogenic sources, compared to a total production of biogenic hydrocarbons of at least 10^{17} tons annually.

Experience of major oil spillages such as that caused by the wreck of the Torrey Canyon in 1967 (Smith, 1968) has shown that the principal danger is damage to the littoral and sublittoral communities when the oil is washed ashore. Even large oil slicks on the open seas readily disperse, a process which may be aided by the use of oil dispersants or other means. The more volatile fractions evaporate, and the heavier fractions eventually sink and are degraded by microbial action. Commonly, substantial amounts of heavier material form tar balls, usually a few millimetres in diameter, which are of more or less neutral buoyancy and float around in the water for long periods. There are few regions of the world where tar balls are not found, and in some areas, such as the Mediterranean, they are washed up on beaches where they form a considerable nuisance. Whether tar balls or sunken oil have any serious biological consequences is not known. Since oil is degraded by micro-organisms, it has been suggested that oil could have a eutrophicating effect similar to that which may occur when other forms of organic material are discharged to water. Most experience of oil spillages relates to warm or temperate seas, and the potential dangers of serious pollution in the more recently-developed arctic oilfields have not been fully assesed. Oil which is washed ashore after major spillages initially causes dramatic damage in the littoral and sublittoral zones, but in most of the major spills which have occurred in the last twenty years it has been found that recolonisation and restoration of the affected areas occurs spontaneously and quite rapidly, within three or four years in high-energy environments (rocky, wave-washed areas) and over a rather longer period in more sheltered shores. Wave action has an important effect in physically breaking up and dispersing the oil; in sheltered areas, oil may initially be simply buried with sand and may be subsequently re-released during rough weather, so the time required for complete recovery may be ten or more years. Although polluted shores recover fairly quickly and spontaneously, the community which is re-established may differ in some ways from that which was destroyed. It is now generally agreed that a large part of the damage associated with oil spillages is in fact attributable to the toxicity of the oil dispersant materials which are often applied as an emergency measure. The newer oil dispersants are much less toxic than those in use formerly, but it is now widely accepted that they should be used cautiously, if at all, in coastal waters. They represent an additional toxic hazard and are usually applied for aesthetic rather than ecological reasons, for example to accelerate the removal of the oil from recreational beaches in tourist areas. Surprisingly little research has been done in areas subject to natural oil seepage from under the sea bed, but the available data indicate that in these areas the effects of the oil are readily detectable in the immediate locality of the discharge but do not spread very far. In zones of intermediate oil concentration, the microbial breakdown of oil may cause an enriching effect leading to increased productivity, if not an actual increase in species diversity.

Chronic oil pollution of coastal areas, in harbours and around oil installations and refineries, is much more significant, in terms of the total area affected, than the more dramatic major accidents. The effects of chronic pollution are readily detectable in such locations, but experience has shown that while some effects may be long-lasting, rapid improvements take place following the installation or upgrading of appropriate pollution control facilities. The effects of offshore oil installations, if they are properly managed, are confined to the immediate vicinity of the installations. There is little evidence that oil pollution in the open seas associated with normal shipping operations has any effects other than purely local ones. Similarly, there is no evidence that oil pollution of any kind has far-reaching consequences for ecologically or commercially important species or communities (Clark, 1982). Sea bird deaths in large numbers are conspicuously related to oil pollution, but while this may be distressing for some people the numbers involved are actually small in relation to natural mortality rates. In any case, adult mortality rates are not, for sea bird populations, generally the principal factor governing population density, and no sea bird population is known to be seriously threatened by oil. No evidence has been found to suggest that fish stocks, except in localised areas, are adversely influenced by oil pollution (or, indeed, pollution of any other kind in the sea). Planktonic organisms are, like polluting oil, mobile and patchily distributed; although they are sensitive to oil, damage again appears to be confined in time and space to the immediate vicinity of the discharge.

It is reasonable to conclude that some of the more extravagant fears about oil pollution of the sea are probably unfounded. Nevertheless oil is a serious pollution threat. Although the problems may be localised, there are many large areas detectably affected. Some habitats of particular ecological importance, such as mangrove swamps, mud flats, coastal bays and estuaries, are permanently at risk from accidents. The potential does exist for a local problem to have far-reaching consequences; for example, a major spill occurring in the nursery grounds of a fish population, if it occurred at an appropriate time, could have major ecological and economic implications. Finally, the view that oil pollution is a matter for local rather than global concern may simply reflect the fact that the techniques and knowledge currently available are not sufficient to allow the detection of potentially important but subtle effects. The threat of oil pollution therefore fully justifies both further research and the continuing efforts to control and abate it.

7.2.2   Sewage and domestic wastes

Coastal towns and cities all over the world discharge sewage, often completely untreated, into the sea. In addition, substantial quantities of sewage enter the sea from grossly-polluted rivers. It is now widely recognised that this has created significant ecological damage and measurable public health risks, particularly in the many locations where sewage is discharged at the shoreline or through short outfalls perhaps only 100 or 200 metres in length. The health hazards arise mainly from the high numbers of pathogenic organisms which may be found in the water and on the beaches, presenting a danger to bathers and recreational users of the beach, and from the accumulation of pathogens in fish and shellfish which may be used for human consumption. Reish (1984) provides a detailed review of the effects of sewage discharges in marine environments.

The effects of sewage in the marine environment are, in principle, exactly the same as the effects in fresh water, and arise through the same mechanisms (see section 2.3). The extent to which the adverse consequences manifest themselves seems to depend greatly on local conditions. In some cases, apparently large discharges appear to be accommodated without serious disturbance to the ecosystem, and the nutrient-enriching effect may even be considered beneficial. In other cases, severe disturbance of the benthic communities, even to complete elimination, has been recorded. In addition, problems similar to those associated with extreme eutrophication in lakes occurs, including deoxygenation of the hypolimnion and sea bed, and the creation of algal blooms. Many of the algal blooms are associated with the production of algal toxins which are damaging to fish and other species of both ecological and commercial significance. As in fresh water, it is possible to rank species of benthic invertebrates in approximate order of susceptibility to organic pollution. The most severely polluted zones contain few species which are apparently tolerant of pollution, and species diversity tends to increase in a fairly predictable fashion as the intensity of pollution decreases. Thus the effects of sewage pollution on the marine benthic community can be monitored in a similar way to the effects in fresh water, at least in some localities. Bellan (1970) described a zonation of benthic species in the vicinity of sewage discharges into the sea off Marseilles, a pattern which has been observed, with some variations, in other locations (Perkins, 1974, 1979). However, in a number of locations where such a pattern may be expected, by reason of known substantial sewage discharges, it is not apparent. This may be due to local conditions of tides and currents which affect sediment movements and the dispersion of the pollution, but until more data are available it is difficult to decide whether the absence of the expected zonation indicates that the pollution is exerting little effect, or whether the available techniques of monitoring are insufficiently sensitive to allow significant effects to be detected. For various reasons, techniques of biological monitoring are less well-developed in the marine environment than in fresh water (see section 7.2.3).

Whether or not sewage discharges exert serious effects on the receiving environment therefore appears to depend upon the extent to which satisfactory dispersion can be achieved through the design and location of the outfall in relation to the local conditions of tides, winds and currents. Determining the optimum strategy for marine sewage disposal requires a detailed knowledge of local conditions, and probably each case needs individual consideration. Frequently, improvements to the local environment can be achieved by constructing a sufficiently long outfall of suitable design and in a favourable location. In many circumstances, however, this is not technically or economically feasible and actual treatment of the sewage is required. In Britain, a common solution is to give sewage primary and sometimes secondary treatment, discharging the liquid effluent through a relatively short outfall (sufficiently long to avoid public nuisance, for example) and dumping the sludge from specially-constructed ships further out to sea. The extent to which this practice is acceptable is currently controversial. There is greater agreement, however, that sewage pollution of the sea is a very serious problem in many areas of the world, and that the waste disposal facilities of coastal towns and cities are in urgent need of upgrading. In Europe, for example, at least three substantial sea areas, totalling up to one million square kilometres in area, are thought to be in imminent danger. These are the northern Adriatic, the Baltic and parts of the eastern North Sea. In these seas, periodic deoxygenation of the sediments and water column have been recorded, often associated with heavy losses of marine life. In addition, excessive plant growth and toxic phytoplankton blooms are increasingly frequent occurrences, and the potential for serious ecological damage and economic loss affecting wide areas is now recognised.

7.2.3 Toxic pollutants

Of the many toxic pollutants which enter the sea, probably the greatest concern is due to the conservative pollutants, such as heavy metals and refractory organic substances like certain pesticides, polychlorinated biphenyl compounds and similar substances. In coastal areas all over the world, the discharge of toxic wastes causes readily-identifiable ecological damage in localised areas. In at least one well-documented case, Minamata Bay in Japan, widespread and serious human illness was caused by consumption of contaminated seafood (Tsubaki & Irukayama, 1977). There is an enormous literature on the effects of toxic pollutants such as metals (Bryan, 1984) and pesticides (Ernst, 1984) in the sea, but perhaps the most intractable problem is to discover whether their effects are confined to the localities in which they are released, or whether more widespread damage is occurring. Conservative pollutants have, at least in theory, the capacity to accumulate in sediments, and in the tissues of living organisms, until they reach concentrations which may be harmful; such effects may not become obvious until many years after the pollutants have been discharged at what may initially appear to be a 'safe' rate.

In the case of heavy metals such as lead, mercury and cadmium which have no known biological function, inputs to the sea from anthropogenic sources approach or exceed the natural rates of input due to weathering of rocks. (Many refractory organic substances are completely synthetic and in the absence of anthropogenic inputs would not be found in the sea at all.) This suggests that study of the biogeochemical cycles of these substances should be a useful approach. If, for example, it could be shown that the levels of mercury in sea water, sediments or living organisms were steadily rising, it could be argued that existing controls on mercury emissions are inadequate and that contamination is more than simply a local problem. However, in practice great uncertainty exists about some basic aspects of the biogeochemistry of mercury and other metals. Aston *et al.* (1986) point out that estimates of the mercury concentrations found in open sea water have actually been lowered by about one order of magnitude over the last two decades. This has happened because the levels being measured are at or close to the detection limits of the analytical procedures. As measurement techniques improved, the problems of contamination during the taking and processing of the sample became more apparent, and as increased precautions against contamination were employed, the levels of mercury recorded have actully declined. This casts considerable doubt on the validity of most of the available data, including many recent measurements which did not make use of the very best available techniques; in effect, there is no realistic baseline with which to compare the values currently being recorded.

An alternative approach is to measure the concentrations of toxic substances in sediments and living organisms, and a large volume of information is being accumulated. Collation of this information into a comprehensive bank of baseline data presents formidable problems of quality control, and difficulties of comparison caused by differences in methodology, but will undoubtedly be useful in the future. Such studies do frequently indicate, however, the existence of 'hot spots', or areas of apparently elevated concentrations of toxic matter, usually associated with urban or industrial pollution. The extent to which these elevated concentrations are harmful can only be assessed by toxicological studies. Bryan (1979) drew attention to some of the difficulties of interpreting data on the measured tissue levels of pollutants in marine organisms (see also section 4.3.3). As with many other forms of toxicological investigation, the large differences between the best available experimental techniques and the reality of the field is particularly large when studying marine pollution. In laboratory experiments, uptake of pollutant occurs largely from solution. In the sea, animals almost certainly obtain most of their body burden of pollutants by ingestion of contaminated food or sediment. It is well known that the route of entry of a poison greatly influences the extent of its toxic effect. Further, some species appear to be able to immobilise and store pollutants within their bodies, and some appear to have developed a measurable degree of tolerance through genetic and/or acclimation mechanisms. Therefore the existence of a given level of a pollutant in the tissues does not necessarily indicate that it is intrinsically harmful.

In earlier chapters the principles of toxicological investigation and their applications to fresh-water pollution were discussed. In fresh water, it frequently occurs that the concentrations of pollutants actually found in the environment are close to those which can be reliably maintained in controlled laboratory experiments. The results of toxicological studies can, cautiously, be applied successfully to many field situations. In the sea, except for the grossest and most localised forms of pollution, the difference between the concentrations found in the environment and those which can be maintained accurately in controlled experiments is much wider, frequently several orders of magnitude. This, together with the generally inadequate understanding of the effects of the marine environment on the chemical form and speciations of toxic substances (Burton, 1979), casts great doubt on the usefulness of conventional toxicological approaches when applied to marine organisms. Some alternative approaches are discussed in the volume edited by Cole (1979); and some of the recent developments in biochemical and genotoxicological techniques referred to in section 4.3.2 may become important in future.

A promising application of biochemical techniques in assessing the response of the mussel *Mytilus edulis* to pollution was described by Bayne *et al.* (1979). Measurement of some biochemical parameters in animals from different natural populations were used to derive a model to allow prediction of growth rates. The predictions agreed well with the observed growth rates of animals in the field. Animals were transplanted from their natural environment to polluted locations, and subsequent measurements of the biochemical indicators showed that they had responded to the polluted conditions in a way which was detectable. These responses could be related to fecundity. Depending upon the degree of specificity and sensitivity which can be obtained, techniques of this sort probably represent a more directly useful approach to problems of measuring toxicity than those which have been found appropriate in fresh waters. Stebbing (1979) argued that instead of carrying out toxicity tests and seeking to compare the results with the environmental concentrations, it may be more useful in the marine situation to use suitably sensitive organisms to bioassay the sea water; subsequent treatment of the sea water to remove pollutants then, if the response of the test organism is abolished, can be used to indicate which pollutants are exerting a measurable biological effect. Stebbing (1979) reported some success with this approach, using the growth and development of the colonial hydroid *Campanularia flexuosa*. This organism appears to be sufficiently sensitive to allow the bioassay of polluted sea water, and has the advantage that variation due to genetic variability can be excluded by the use of a single clone. (American readers may be confused by the distinction between the terms 'toxicity test' and 'bioassay', since many American authors have in recent years begun to use the two terms synonymously. The difference between the terms is important, as the present example indicates, and has been clearly explained by Brown (1973). It is unfortunate that the distinction in meaning between these terms which was universally accepted for decades has become blurred.)

To the extent that studying the toxicity of pollutants to individual marine organisms is of limited value, the study of toxic effects at the population or community level of organisation is of even greater importance in the sea than in fresh waters. Several attempts have been made using artificial or semi-natural ecosystems (Davies & Gamble, 1979; Steele, 1979), but these are difficult to maintain and control on a sufficiently large scale and with a sufficient degree of replication to allow them to be widely used as predictive tools. In any particular set of circumstances, such experimental systems indicate the range of possible outcomes or mechanisms rather than what actually does happen in the field. They are probably best regarded as attempts to study natural processes on a larger scale, though not necessarily in a more realistic manner, than is possible in the laboratory. The most ambitious experimental ecosystems involve the enclosure of large volumes of water, together with the biomass they contain. However, the very existence of an artificial boundary is likely to interfere with the processes which occur within the enclosure, for example by preventing lateral water movement and by affording a surface for colonisation by algae and micro-organisms which would normally not be present. Boundary effects can, of course, be reduced by increasing the size of the enclosure, but the expense of maintaining the systems soon becomes prohibitive and renders replication of experiments impossible. In practice, it is sometimes found that variability between control replicates exceeds that due to the experimental manipulation. Further, in an enclosure which is sufficiently large to ignore boundary effects, the degree of control available to the experimenter is scarcely better than that available in the completely natural environment; and in any case, in very large enclosures the degree of variability found in different regions within the system is such that the procedures of sampling and analysis which must be employed, and the precision of the results obtained, show little improvement over what would be required if the enclosure were not there at all. Steele(1979) concluded that 'the experiments conducted with large scale ecosystems so far have probably taught us more about the general ecological interactions within such systems than about subtle long-term effects of pollutants'. Such studies have, however, drawn attention to the importance of specific processes which undoubtedly require further study. These include interspecific and intraspecific interactions, and the role of sedimentary processes and of sediment–water interactions in the behaviour of pollutants in ecosystems.

7.3   MARINE BIOLOGICAL MONITORING

Since laboratory experiments and other simulations of the marine environment are inevitably much further removed from reality than is the case with fresh-water systems, the importance of studying the effects of pollution directly in the field is obvious. In Chapter 3, the concept and practice of biological monitoring of polluted waters were discussed. The same principles should, in theory, be applicable in the marine environment, but for various reasons this is not, in practice, always the case. One important general reason is that, apart from severely-polluted coastal areas and other special circumstances, the extent of the community response that may occur as a result of pollution in the sea is probably much smaller in relation to naturally-occurring spatial and temporal variations than is the case in fresh waters. Thus the difficulty of distinguishing pollution effects from natural phenomena is much greater. Perkins (1979) pointed out that even in fresh waters, biological monitoring has been much better developed and more successful in rivers than in lakes; to some extent lakes and the seas resemble one another more closely than they do rivers, particularly in certain physical features. It is, therefore important to consider some of the special problems which may be encountered in seeking to apply conventional techniques of biological monitoring in the marine environment.

7.3.1   Indicator organisms

The concept of indicator organisms is central to biological monitoring (Chapter 3). Two requirements of indicator organisms are relatively limited mobility, and reasonably long life cycles. Thus the organisms will reflect, by their presence, absence or abundance, the environmental conditions of the place where they are situated over their entire life up to the time of sampling. However, many marine organisms have a planktonic phase in the life cycle, so the distribution and abundance of the adults may reflect events which took place at some previous time in a distant location, perhaps at a crucial early stage of the life cycle, or at metamorphosis. Most fresh-water invertebrates have more or less annual life cycles, though some have two or more generations per year and some have overlapping generations in a life-span of two years or so. Perhaps for this reason (and perhaps because depopulated stretches of river can be rapidly recolonised by downstream drift of organisms), in rivers even after catastrophic pollution incidents the 'normal' flora and fauna tend to become re-established within a few months. In contrast, many marine species appear to have much longer life spans, and there is evidence that in marine communities a single event can have profound consequences which, because of the long life-span of the organisms and the nature of interspecific interactions, can persist for a very long time. Lewis (1972) followed population changes of four littoral species on an unpolluted English shore over four years. A combination of climatic events and biotic interactions resulted in repeated large fluctuations in population density, in some cases of more than tenfold. A major determinant of the marked community changes which occurred during the observation period was the success or failure of the planktonic stages in settling and metamorphosis to the sessile adult. Perkins (1979) refers to other examples where single events, not related to pollution, had consequences which persisted for several years. Thus it appears that in marine communities single events, often of natural origin, can cause profound changes which persist long after the disturbing influence has ceased. To this extent, the use of marine organisms as indicators of environmental quality may give a very misleading picture; certainly it is difficult to distinguish pollution-induced changes from natural phenomena.

Many marine species are also known to undergo long-term fluctuations in their levels of abundance. This is particularly well known in fishes, occurs independently of fishing pressure, and completely negates any possibility, on the basis of current knowledge, of reliably using fishery statistics as indicators of marine pollution (Clark, 1986), although the idea is remarkably common. Invertebrates almost certainly undergo similar fluctuations. A cautionary example is *Acanthaster planci*, the 'Crown of Thorns' starfish, which attracted much attention in the Pacific 10–15 years ago. Its increased abundance and damaging effect on coral reef ecosystems was widely attributed to pollution or other anthropogenic influence, but it is likely that in fact periodic increases in its population density are a natural feature of its ecology (Moore, 1978).

This does not mean that the concept of indicator organisms is invalid or inappropriate in marine systems. There are some species, such as the polychaete *Capitella capitata*, which have been widely associated with polluted conditions (Reish, 1973). However, it is clear that the concept cannot simply be transferred from fresh water without further fundamental research. The fact that the idea works successfully in rivers may be due to a fortuitous peculiarity of the river ecosystem — perhaps its simplicity — but it is certainly in part due also to the much greater knowledge that exists about rivers and their biota. In interpreting the data from biological surveys, we implicitly or explicitly compare what we have found with what we expected to find. Our expectations depend upon knowledge of the previous condition of the habitat under investigation, on contemporary knowledge of physically-similar and geographically-adjacent habitats, or at worst upon a broad but detailed knowledge of the ecophysiology of our indicator organisms. For marine organisms, this knowledge does not exist, and is urgently required. Perkins (1979) refers to an example which illustrates the need for more background knowledge of the ecology of marine indicator organisms. Observations of the benthic community along a transect away from a marine waste outfall showed that the diversity and abundance of invertebrates increased with increasing distance from the pollution source. The obvious conclusion, that the outfall was exerting a significant polluting effect, was however erroneous. Subsequent studies in an adjacent, physically-similar area showed that the impoverished community near the pollution discharge was due to natural, physical disturbances caused by wave action, and that the pollution in fact had no detectable effect.

Clearly, then, the concept of indicator organisms in marine pollution studies is likely to remain of limited value until a great deal of additional information is obtained. Survey data will have to be evaluated in the light of the normal, natural variations in community structure and population density which occur in relation to temporal and spatial variations of the physical environment, and due to processes intrinsic to the populations and communities involved. Long-term studies, extending ideally over decades rather than years, of both polluted and unpolluted habitats, are ultimately required before indicator organisms can be used in the marine environment in the same way that they can be used in rivers. Again, however, there is a case to be made that the identification and evaluation of indicator organisms in the marine environment should not necessarily follow the same route which was pursued in fresh-water studies. A very promising and potentially rapid method of identifying marine indicator organisms has been suggested (Gray & Pearson, 1982; Pearson *et al.*, 1983) based upon the study of the distribution of individuals between species in marine communities.

In many ecological communities, a few species are represented by a large number of individuals, while the majority of species are represented by a very small number of individuals. In practice, it is frequently found that in a reasonably representative sample of the community, most of the species are recorded as one or two individuals, one or two species contain nearly all the individuals, and some species are represented by intermediate numbers of individuals. This pattern is also commonly found in the benthic invertebrate communities of unpolluted marine habitats. The species present in the sample may be divided into abundance classes — Class I for species represented by just one individual, class II for those with two or three individuals, class III for four to seven individuals, class IV for eight to 15, and so on. A graph or histogram of the percentage of species belonging to a particular abundance class, against the abundance class (Fig. 7.3) approximates to one half of a normal curve. In

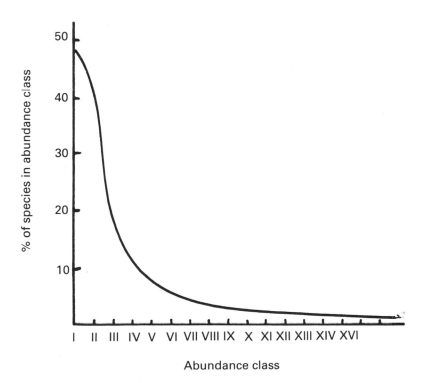

Fig. 7.3 — Log-normal distribution of individuals between species in a typical benthic community. Deviations from the log-normal can be detected by simple statistical tests.

such communities, the distribution of individuals between species is log-normal. In a community which is disturbed by pollution, some species will be rarer and some commoner than might be expected in an unpolluted area, and this log-normal distribution will be distorted. These distortions have been shown to be associated with polluted conditions (Gray, 1979). Pearson *et al.* (1983) argue that species which may be useful indicators of community responses to pollution can be identified objectively, without any need for detailed knowledge of their general ecology or sensitivity to pollution. Species in abundance classes I and II, the rare species, are rejected as indicators. Some of these may be rare because of pollution, but since the majority of species are rare and there are many possible reasons to account for rarity, rarity alone is an insufficient ground for recognising an indicator. The very few species which are found in the higher abundance classes, that is the most abundant species, are also rejected as indicators, for reasons explained below. Those species in abundance classes V and VI are those which are tentatively identified as those which will be the most sensitive indicators in a community of externally-induced disturbance such as pollution. In practice, these species would typically be those which contain between 16 and 63 individuals per sampling unit; the sampling unit being typically five replicate $0.1 \, m^2$ grab samples (Pearson *et al.* 1983).

It is implicit in this method that the conventional approach to the recognition of indicator species is rejected. To take an example, the polychaete *Capitella capitata* is, conventionally, widely regarded as an indicator of organic pollution in marine habitats, because it is often found in abundance in such conditions. Gray (1979) and Pearson *et al.* (1983) argue that this view is mistaken. Firstly it appears that *C. capitata* is not, in fact, particularly tolerant of pollution. Several other polychaete species are measurably more resistant to various forms of pollution in experimental studies, but are not found abundantly in polluted conditions in the field. They argue that the association of *C. capitata* with polluted conditions is not due to its tolerance of pollution, but to the fact that it is an opportunist species with a high reproductive capacity and good powers of dispersal, which allow it readily to colonise disturbed areas, and quickly to make good any population losses due to adverse environmental conditions. Its population density is subject, in any particular location, to rapid and massive fluctuations. Secondly, the authors argue that *C. capitata* is a complex of at least six subspecies, which are difficult to distinguish by conventional taxonomic means. Similar considerations apply to many of the marine species which are regarded, on the conventional approach, as useful indicators. Therefore the use of the most abundant species as indicators is potentially misleading.

In contrast, the use of this novel approach offers several interesting possiblities. Firstly, the deviation of community structure from the log-normal is amenable to statistical tests of significance, which pollution and diversity indices are not. Secondly, it allows the identification, by objective means, of a relatively small number of species which, in any particular location, are likely to respond to environmental change. This allows the expense and effort of monitoring to be concentrated on these species, rather than on more abundant, more taxonomically-difficult species whose significance as indicators is in any case in doubt. In their analysis of six extensive data sets from different parts of the world, Pearson *et al.* (1983) found that their method rapidly identified similar lists of species to those determined by more time-consuming methods. Further, the method allows the identification of indicator species specific to particular forms of pollution. Their analysis also suggests that in different parts of the world, different indicator species will be involved in community responses to pollution. Although, therefore, it may be necessary to identify different sets of indicator species for each form of pollution and each individual location, the method affords the means whereby this can be done relatively quickly.

7.3.2   Sampling methodology

The importance of adequate sampling techniques, and the relationship between sampling strategy, data analysis and the validity of the final results were discussed in Chapter 3. The same considerations of course apply also to the marine environment. Unfortunately, sampling in the marine environment is notoriously difficult and expensive, especially if the use of offshore vessels is involved. There is therefore an inevitable tendency to attempt to minimise the sampling effort undertaken in marine survey work. There is a danger that in some cases the effort undertaken is, in the end, wasted; or worse, that inaccurate information is obtained which serves only to confuse or mislead. This may appear, to some, an extreme statement; but as recently as 1982 the editor of a well-known journal in the field publically drew attention to the poor quality of much research, and cited inadequate sampling as a major reason (Clark, 1982b). Hartley (1982) discussed some common failings of many investigations, and made some specific recommendations. In particular, he showed that the number of samples required to estimate accurately the population density of benthic invertebrates is, as in fresh water, often inconveniently large (Table 7.1). Other sources of error, and of difficulty in comparing results from different sources, are the different efficiencies of the various designs of grab (or other sampling device) which are used, and variations in the procedure employed to extract the animals, usually by some form of sieving. Sources of guidance on the subject of sampling in marine ecological research are available (e.g. Dybern *et al.*, 1976; Stirn, 1981; Holme & McIntyre, 1984), but an urgent need is for wider general agreement on standardised methods for pollution surveys.

7.3.3   Data analysis

The interpretation of biological survey data is essentially a series of comparisons — temporal, spatial or both — and some of the methods available were described in Chapter 3. One difference between marine survey data and that from fresh water is immediately apparent. For practical purposes, a river is usually considered as a one-dimensional system, that is having length, but negligible width or depth. At least, width and depth effects are, in practice, usually ignored in biological surveys; and the comparison of communities above and below a source of pollution, and at various distances from the pollution source usually allow any effects of pollution on the receiving water fauna to be readily detected. The relatively strong and unidirectional flow of rivers, and the accretion of unpolluted water along the length of the river, tend to accentuate differences between communities at different points. Marine habitats, in contrast, are at least two-dimensional, and may be three dimensional if the pelagic biota are to be considered. Further, since observations at a single point in time are clearly of very limited value, the additional dimension of time is likely to be involved in many surveys. Thus the analysis of marine survey data is intrinsically

Table 7.1 — Comparison of the number (n) of replicate 0.1 m^2 samples required to give standard error of counts per taxon equal to 20% of the mean. Data from survey of Forties oilfield (Hartley, 1979), given by Hartley (1982). The number of samples required (n) is generally <5 for gross taxonomic groupings (polychaetes, molluscs, crustaceans). However, for individual species the number of replicate samples required is much larger. Note that only the more abundant species have been considered. For rare species, the number of samples required may be much larger

	X	$\overline{\text{X}}$	s^2	n	Range and mean of values of (n) for numerically dominant species[a]	
					Range	Mean
Station 15						
Polychaetes	420	84.0	278.5	0.99		
Crustaceans	58	11.6	17.3	3.21	0.6–18.6	5.51
Molluscs	132	26.4	11.3	0.41		
Station 16						
Polychaetes	407	81.4	419.3	1.58		
Crustaceans	59	11.8	11.2	2.01		
Molluscs	157	31.4	31.3	0.79		
Station 17						
Polychaetes	431	86.2	237.7	0.92		
Crustaceans	66	13.2	16.2	2.32	2.7–96.1	20.68
Station 18						
Polychaetes	341	68.2	17.2	0.09		
Crustaceans	62	12.4	32.3	5.25	1.5–14.9	6.00
Molluscs	127	25.4	43.3	1.68		

[a] Regarded as those top ranked species contributing to the first 50% of the number of individuals at each station (10–12 taxa for these stations).

more complex than that of river data. It follows that the data required from a marine survey, for the purposes of meaningful spatial and temporal comparisons, the design of the sampling programme which produces the data, and the methods of data analysis and interpretation employed, must also be more complex.

Index methods (section 3.4) have often been found useful in analysing survey data in fresh-water surveys, and have also been applied to marine work. Diversity indices have been widely used, but are subject to the limitations discussed in section 3.4. (See also Reish, 1984; Gray & Pearson, 1982.) The use of deviation from the log-normal distribution of individuals among species to identify disturbed communities (Gray & Mirza, 1979; see also section 7.3.1) seems to offer some advantages. Multivariate methods of analysis such as cluster analysis (section 3.4) appear promising. Analysis based on inspection of a matrix of similarity coefficients, but stopping short of constructing the dendrogram, was recommended under the name 'trellis method' by Stirn (1981). Some examples of cluster analysis applied to the interpretation of marine survey data are discussed by Reish (1984). Biotic and pollution indices have been rarely used, for the obvious reason that this form of index relies on some form of ranking of known species in order of susceptibility to pollution, and for most marine species it is as yet impossible to do this. Perhaps the nearest approach to the idea is the nematode/copepod ratio. It has frequently been observed that in samples of the benthic meiofauna, the ratio of nematodes to copepods varies with the intensity of organic pollution in the area from which the samples were taken. The use of this ratio as an index of organic pollution was suggested by Rafaelli & Mason (1981). The idea has not been fully tested. Amjad & Gray (1983) found that it produced satisfactory results in Oslo Fjord, but in the Firth of Forth (Shiells & Anderson, 1985) the ratio varied along apparent pollution gradients in an inconsistent manner

This latter example illustrates an important difference between maritime habitats and rivers which is often overlooked. Whereas, in a river, it is reasonable to expect that a pollution gradient exists along the length of the river, with pollution decreasing with increasing distance downstream from the source of the discharge, in estuarine and marine habitats this is not the case. In these environments there are complex vertical and horizontal patterns of water movement, which are not constant or unidirectional and which are influenced greatly by tides and winds. Therefore it cannot be assumed that pollution decreases unidirectionally with increasing distance from the point of discharge. In the above case, it is possible either that the pollution index used is not valid, or that the pollution gradient is not as it is assumed to be, or indeed that the pollution is not exerting a sufficiently large effect to be detectable.

7.4 INTERNATIONAL CO-OPERATION ON MARINE POLLUTION

The control and prevention of marine pollution very obviously requires international co-operation, and it is relevant, by way of conclusion, to look at some of the steps which have been taken to achieve this objective. In many parts of the world, coastal states have agreed by treaty to co-operate in various ways to monitor, ameliorate or eliminate the effects of pollution of the seas. Examples include the North Sea and north-east Atlantic areas, which are governed by the Oslo and Paris conventions. The scope and effectiveness of these agreements have been reviewed by Bjerre & Hayward (1984). Inevitably, progress is to some extent limited by differences between the various technico-legal approaches which may be adopted in individual states, quite apart from any purely economic aspects of national self-interest. A typical problem is that Britain, for example, dumps substantial quantities of sewage sludge at sea, because most of its large coastal cities have sewage treatment facilities which are designed to protect estuaries and coastal waters from pollution, but which generate large amounts of sludge. Some other countries do not dump sludge at sea, but this is at least in part due to the fact that they have not constructed the sewage treatment facilities which generate the sludge! The question of which waste disposal strategy creates the greater environmental hazard is capable of generating both scientific and political controversy.

In other parts of the world, some of the most vulnerable seas are bounded by coastal states which are underdeveloped, short of economic and human resources and struggling with numerous technical and socio-economic difficulties which readily assume priority over concern for the environment. The need of such countries for assistance was one of the reasons for the creation of the United Nations Environment Progranmnme (UNEP), as one of the far-reaching consequences of the UN Conference on the Human Environment in Stockholm during 1972. One of UNEP's major programmes, the Mediterranean Action Plan, may be taken as a useful example of international co-operation to combat marine pollution.

7.4.1   The Mediterranean Action Plan   (— solo)

The Mediterranean is a unique sea. It supports at least 100 species which are not found elsewhere. Its coast is home for about 100 million people, and is visited annually by about 30 million tourists. It is one of the world's busiest shipping areas, and a large proportion of the population of its coastal states depend upon it, directly or indirectly, for their food and livelihood. Some of the coastal states are wealthy, industrialised countries and others are rapidly undergoing development and industrialisation; still others are among the poorest and most backward countries in the world. At least two countries, Libya and Algeria, are major oil producers, and the search for offshore oil and minerals in the Mediterranean is well advanced. Politically, the Mediterranean states vary from Western-style democracies to military dictatorships. Long-standing antagonisms between some states, such as Greece and Turkey, Israel and the Arab states, are potential impediments to agreement. The narrow coastal strip is under great pressure from industrial development, population drift from the poorer and climatically harsher hinterlands, and developments associated with tourism. The sea is more or less enclosed. The fresh-water input from rivers is considerably less than the rate of evaporation, so the salinity of the Mediterranean is measurably higher than that of most seas. The loss of water by evaporation is compensated for by the influx of cold Atlantic water through the straits of Gibraltar; this influx displaces a smaller volume of warm water which flows out at the surface in the opposite direction. There is no significant interchange of water with the Black Sea, and it is estimated that the volume of the Mediterranean is replaced only once in every eighty years. The sea receives sewage and industrial wastes from 120 coastal cities, about 80% of it untreated or inadequately treated. Identifiable health hazards and/or ecological damage affect 25% of the Mediterranean coastline. In addition, the sea receives 25% of the total amount of oil discharged to the world's seas, excluding major accidents. Table 7.2 lists UNEP's estimate of the quantities of some common pollutants discharged annually to the Mediterranean. It is not yet known whether the sea as a whole is under serious threat, but the existence of very severe local problems in many areas is beyond doubt.

For these reasons, one of UNEP's first activities was the development and implementation of the Mediterranean Action Plan (MAP). This plan became the prototype for a number of similar schemes which now cover, among other areas, the coastal seas of South America, the Caribbean, and most of the coastal seas of West Africa, the Indian Ocean and South-East Asia. In 1975 UNEP convened a conference of Mediterranean coastal states which approved the MAP. The plan called for the adoption of a series of treaties, the creation of a pollution monitoring and research network, and a socio-economic programme to reconcile development priorities with environmental protection. The Barcelona Convention (United

Table 7.2 — Estimates of the quantities of pollutants entering the Mediterranean Sea annually from anthropogenic sources. Data from UNEP (1985)

Pollutant	Quantity (tons)
Mineral oils	120 000
Phenols	12 000
Synthetic detergents	60 000
Mercury	100
Lead	3 800
Chromium	2 400
Zinc	21 000
Phosphorus	320 000
Nitrogen	800 000

Nations, 1978), adopted in 1976 by all the Mediterranean coastal states (except Albania) and the EEC, binds the participants to co-operate in four major areas; integrated planning, pollution monitoring and research, legal matters, and institutional and financial arrangements. The United Nations Environment Programme was appointed to carry out co-ordinating and secretariat functions. The costs of the plan are mainly met by contributions in kind (manpower, laboratory facilities) from institutions of the participating states; only EEC and some of the wealthier states make substantial cash contributions. The UNEP itself has funds allocated to it by the UN. The money pays for the secretariat functions, scientific conferences, research grants and the supply of capital equipment to the poorer laboratories.

All MAP activities are derived from the Barcelona Convention and its protocols. The 1976 Protocol on the Prevention of Pollution by Dumping from Ships and Aircraft (United Nations, 1978) prohibits the disposal at sea of a 'black list' of substances such as mercury, cadmium and radioactive waste. A second, 'grey list' of less noxious substances is closely regulated. The 1980 Protocol for the Prevention of Pollution from Land-Based Sources (United Nations, 1980) explicitly recognises that ultimately, marine pollution is derived from land-based sources by discharge from rivers, land runoff and atmospheric fallout. It binds the states to take technical and legislative measures to control sewage, industrial wastes and agricultural chemicals at source. It also has 'black' and 'grey' lists. The 1976 Protocol concerning Co-operation in Combating Pollution by Oil and Other Harmful Substances in Cases of Emergency (United Nations, 1978) commits the governments to co-operate in the case of serious accidents. A regional oil-combating centre in Malta provides training, technical advice and equipment, formulates contingency plans and co-ordinates the responses of governments to accidents. The 1982 Protocol concerning Mediterranean Specially Protected Areas (United Nations, 1984) agrees policies for the designation and protection of the habitats of endangered species and other areas of biological importance. About 40 such areas have been designated so far, and this activity is co-ordinated from a regional centre in Tunis.

The Blue Plan, the Priority Actions Programme and MEDPOL derive from the Barcelona Convention itself. The Blue Plan, directed from a regional centre in Sophia Antinopolis, France, is concerned with long-term planning. Pressure on the sea mounts through the development of industry and tourism, and population movements towards the limited coastal strip. The Blue Plan reports on subjects such as water resources, population trends, tourism, food requirements and transport. It offers advice to governments on the local, national, or regional consequences of particular development policies. Its aim is to reduce long-term pressures on the sea by providing a rational basis for environmentally-acceptable development, and to advise on policies to prevent or reverse harmful development. The Priority Actions Programme offers technical advice and financial support for specific projects which are environmentally advantageous and can be implemented fairly swiftly. Directed from a regional centre in Split, Yugoslavia, its activities include projects on aquaculture, solar energy, water resource development, waste disposal and land use planning in earthquake zones.

The Programme for Pollution Monitoring and Research in the Mediterranean (MEDPOL) is directed from Athens, which is also the overall co-ordinating office for the entire MAP. Phase I (1975–1980) began a monitoring programme and assessed research needs. Harmonised monitoring programmes were devised for pollutants in water, sediments and organisms. Reference methods were devised for chemical and biological sampling and analysis, and tested by intercalibration exercises. Laboratories were equipped and trained in the appropriate techniques. A mobile team of instrument engineers was established to ensure adequate maintenance of equipment. Preliminary studies were undertaken on, for example, the biogeochemical cycles of selected pollutants; input of pollutants from the atmosphere; and the microbiological quality of coastal waters. Phase II of MEDPOL continues the monitoring activity, and has a research component which is divided into the 12 topic areas shown in Table 7.3. Periodically, scientists working on each topic meet to review progress and identify research needs. Up to the end of 1985, expenditure on the MAP was approximately US $25 million.

Despite the existence of this ambitious and comprehensive framework, it is probably true that the actual achievements of the MAP so far, though significant, are modest in relation to the plan's potential. This may be disappointing, but it is hardly surprising. Some member states, as happens with all international agreements, pursue their treaty obligations more assiduously than others. A major problem remains the shortage of suitably qualified and experienced personnel, aggravated by the limitations of the educational system, and its shortage of resources, in many countries. This has a number of consequences. Firstly, a great deal of the monitoring and research data obtained is in practice of doubtful value. Secondly, it is arguable that some substantial part of the total research effort is wasted on relatively unimportant problems of purely local relevance, or is based on conceptual and technical approaches which are outdated. This occurs, in part, because within the MAP the overall direction of the research is strongly influenced by the scientists from the region who are participating in it. Even senior staff from many countries have, until recently, had limited access to the international literature and have had little opportunity to experience fully the pace and rigours of modern research. Also, precisely because skilled staff are scarce, they are often under pressure to concen-

Table 7.3 — The research activities of the MEDPOL component of the Mediterranean Action Plan

---

ACTIVITY A: development and testing of sampling and analytical techniques for monitoring of marine pollutants;

ACTIVITY B: development of reporting formats required according to the Dumping, Emergency and Land-Based Sources Protocols;

ACTIVITY C: formulation of the scientific rationale for Mediterranean Environmental Quality Criteria;

ACTIVITY D: epidemiological studies related to Environmental Quality Criteria;

ACTIVITY E: guidelines and criteria for the application of the Land-Based Sources Protocol;

ACTIVITY F: research on oceanographic processes;

ACTIVITY G: research on toxicity, persistence, bioaccumulation, carcinogenicity and mutagenicity;

ACTIVITY H: eutrophication and concomitant plankton blooms;

ACTIVITY I: pollution-induced ecosystem modifications;

ACTIVITY J: effects of thermal discharges on coastal organisms and ecosystem;

ACTIVITY K: biogeochemical cycles of specific pollutants;

ACTIVITY L: pollutant transfer processes.

At the end of 1984 there were 102 on-going research projects carried out by 62 centres in 16 coastal States.

trate their efforts on local problems which need urgent attention, rather than on more fundamental long-term research which may ultimately be more beneficial. These problems are endemic in many parts of the world, and perhaps one of the major achievements of the MAP and similar schemes will be the emergence of a corps of scientists, suitably experienced in modern techniques and approaches by virtue of the opportunity they have had to participate more fully in the international scientific community. To this can be added other successes. The pollution threat to the Mediterranean has been identified and at least partially quantified, and measures are in hand to combat the threat at the technical and legislative levels. Practical conservation measures have been implemented. Cheap and unbiased technical advice is now available to all the countries in the region, and many cities and industries have begun to construct pollution control facilities for the first time.

In the preceding chapters, which were mainly concerned with fresh-water pollution, it became clear that enormous efforts and ingenuity are required to preserve and manage fresh-water resources in the face of the many pressures upon them. Those efforts have, by most measures, been enormously successful: the pollution pressures have increased, but the waters have not proportionately deteriorated, and many are distinctly improved in comparison with the situation of, say, fifty or one hundred years ago. We use more water, of higher quality, than ever before; we dispose of more waste than ever before; but outbreaks of waterborne diseases are, at least in developed countries, events which attract national attention and comment rather than a dismal fact of everyday life and the commonest cause of death. Many problems remain, and new ones will arise, but we can be reasonably certain that, whatever they turn out to be, there will be some basis available for tackling them. The problems of marine pollution appear, at present, more intractable. We do not, in fact, yet know what the most serious problems are; and we cannot be sure that the means are yet available for solving them. Most of the work to be done still lies ahead.

References

Abel, P. D. (1974) Toxicity of synthetic detergents to fish and aquatic invertebrates. *J. Fish. Biol.* **6** 279–298

Abel, P. D. (1976) Toxic action of several lethal concentrations of an anionic detergent on the gills of the brown trout, *Salmo trutta* L. *J. Fish. Biol.* **9** 441–446

Abel, P. D. (1978) Mortality patterns in trout exposed to an anionic detergent in relation to concentrations and mechanisms of toxic action. *Freshwater Biol.* **8** 497–503.

Abel, P. D. (1980a) Toxicity of γ-hexachlorocyclohexane (Lindane) to *Gammarus pulex*: mortality in relation to concentration and duration of exposure. *Freshwater Biol.* **10** 251–259

Abel, P. D. (1980b) A new method for assessing the lethal impact of short-term, high-level discharges of pollutants on aquatic animals. *Progr. Wat. Tech.* **13** 347–352

Abel, P. D. (1988) Pollutant toxicity to aquatic animals — methods of study and their applications. *Rev. Environ. Health,* **8**, (in press).

Abel, P. D. & Garner, S. M. (1986) Further comparisons of median survival times and median lethal exposure times for *Gammarus pulex* exposed to cadmium, Permethrin and cyanide. *Water Res.* **20** 579–582

Abel, P. D. & Green, D. W. J. (1981) Ecological and toxicological studies on the invertebrate fauna of two rivers in the Northern Pennine orefield. In: Say, P. J. & Whitton, B. A. (eds) *Heavy metals in Northern England — environmental and biological aspects.* University of Durham, Department of Botany

Abel, P. D. & Papoutsoglou, S. E. (1986) Lethal toxicity of cadmium to *Cyprinus carpio* and *Tilapia aurea. Bull. Environ. Contam. Toxicol.* **37** 382–386

Abel, P. D. & Skidmore, J. F. (1975) Toxic effects of an anionic detergent on the gills of rainbow trout. *Water Res.* **9** 759–765

Abram, F. S. H. (1973) Apparatus for control of poison concentration in toxicity studies with fish. *Wat. Res.* **7** 1875–1879

Addison, R. F. (1984) Hepatic mixed function oxidase (MFO) induction in fish as a possible biological monitoring system. In: Cairns, V. W., Hodson, R. V. & Nriagu, J. O. (eds) *Contaminant Effects on Fisheries*. John Wiley, New York

Adelman, I. R. & Smith, L. L. (1972) Toxicity of hydrogen sulphide to goldfish (*Carassius auratus*) as influenced by temperature, oxygen and bioassay techniques. *J. Fish. Res. Bd Can.* **29** 1309–1317

Adelman, I. R., Smith, L. L. & Siesenopp, G. D. (1976) Effect of size or age of goldfish and fathead minnows on use of pentachlorophenol as a reference toxicant. *Water Res.* **10** 685–687

Alabaster, J. S. (1972) Suspended solids and fisheries. *Proc. R. Soc. Lond.* **B180** 395–406

Alabaster, J. S., Garland, J. H. N., Hart, I. C. & Solbé, J. F. de L. G. (1972) An approach to the problem of pollution and fisheries. *Symp. Zool. Soc. Lond.* No. 29, pp. 87–114

Alabaster, J. S. & Lloyd, R. (eds) (1980) *Water quality criteria for freshwater fish.* Butterworths for the Food and Agriculture Organization, London

Alexander, R. McN. (1970) *Functional design in fishes.* 2nd edn. Hutchinson, London

American Society for Testing and Materials (1973) *Biological methods for the assessment of water quality*. Special Technical Publication No. 528. Philadelphia

American Public Health Association (1975) *Standard methods for the examination of water and wastewater.* 14th edn. Washington

American Public Health Association (1981) Standard methods for the examination of waters and wastewaters. 15th edn. American Public Health Association, American Waterworks Association, American Water Pollution Control Federation, Washington

Amjad, S. & Gray, J. S. (1983) Use of the nematode-copepod ratio as an index of organic pollution. *Mar. Poll. Bull.* **14** 178–181

Armitage, P. D. (1980) The effects of mine drainage and organic enrichment on benthos in the river Nent system, Northern Pennines. *Hydrobiologia* **74** 119–128

Arthur, D. R. (1975) Constraints on the fauna in estuaries. In: Whitton, B. A. (ed) *River ecology* Blackwell Scientific Publications, Oxford

Arthur, J. W., Zische, J. A. & Erickson, G. L. (1982) Effect of elevated water temperature on macroinvertebrate communities in outdoor experimental channels. *Water Res.* **16** 1465–1477

Ash, I., McKendrick, G. D. W., Robertson, M. H. & Hughes, H. L. (1964) Outbreak of typhoid fever connected with corned beef. *Br. Med. J.* **1** 1474–1478

Aston, S. R., Fowler, S. W. & Whitehead, N. (1986) Mercury biogeochemistry in the Mediterranean marine environment: an assessment of contamination. FAO/UNEP/WHO/IOC/IAEA Meeting on the biogeochemical cycle of mercury in the Mediterranean. Sienna, August 1984. FAO Fish. Rep. 325 (Supplement) pp. 8–19.

Baker, J. P. & Schofield, C. L. (1985) Acidification impacts on fish populations: a review. In: Adams, D. D. & Page, W. P. (eds) *Acid deposition: environmental, economic and policy issues*. Plenum Press, New York

Ball, I. R. (1967a) The relative susceptibilities of some species of freshwater fish to poisons. I. Ammonia. *Wat. Res.* **1** 767–775

Ball, I. R. (1967b) The toxicity of cadmium to rainbow trout (*Salmo gairdneri* Richardson). *Water Res.* **1** 805–806

Ball, I. R. (1967c) The relative susceptibilities of some species of freshwater fish to poisons. II. Zinc. *Water Res.* **1** 777–783

Ballard, J. A. & Oliff, W. I. (1969) A rapid method for measuring the acute toxicity of dissolved materials to marine fishes. *Wat. Res.* **3** 313–333

Balloch, D., Davies, C. E. & Jones, F. H. (1976) Biological assessment of water quality in three British rivers: the North Esk (Scotland), the Ivel (England) and the Taf (Wales). *Wat. Pollut. Control* **75** 92–110

Barnes, R. S. K. (1984) *Estuarine biology* 2nd edn. Edward Arnold, London

Barnes, R. S. K. & Green, J. (eds) (1972) *The estuarine environment.* Applied Science Publishers, London

Bayne, B. L., Moore, M. N., Widdows, J., Livingstone, D. R. & Salkeld, P. (1979) Measurement of the responses of individuals to environmental stress and pollution: studies with bivalve molluscs. *Phil. Trans. R. Soc. Lond.* **B286** 563–581

Bell, H. L. (1971) Effect of low pH on the survival and emergence of aquatic insects. *Water Res.* **5** 313–319

Bellan, G. (1970) Pollution by sewage in Marseilles. *Mar. Pollut. Bull.* **1** 59–60

Benes, V. (1978) Toxicological aspects of the water we drink. In: Plaa, G. L. & Duncan, W. A. M. (eds) *Proceedings of the First International Congress on Toxicology.* Academic Press, New York

Benenson, A. S. (1982) Cholera. In: Evans, A. S. & Feldman, H. A. (eds) *Bacterial infections of humans.* Plenum Publishing Corporation, New York

Bjerre, F. & Hayward, P. A. (1984) The role and activities of the Oslo and Paris Commissions. In: Lack, T. J. (ed) *Environmental protection: standards, compliance and costs.* Ellis Horwood for Water Research Centre, Chichester

Bliss, C. I. (1935) The calculation of the dosage–mortality curve. *Ann. appl. Biol.* **22** 134–167.

Bliss, C. I. (1937) The calculation of the time-mortality curve. *Ann. appl. Biol.* **24** 815–852

Block, J.-C. (1983) Viruses in environmental waters. In: Berg, G. (ed) *Viral pollution of the environment.* CRC Press, Boca Raton

Bond, R. G. & Straub, C. P. (eds) (1974) *Handbook of environmental control Vol. IV. Wastewater treatment and disposal.* CRC Press, Cleveland

Bonde, D. J. (1977) Bacterial indication of water pollution. *Adv. aquat. microbiol.* **1** 273–364

Brillouin, L. (1951) Maxwell's demon cannot operate: Information and entropy I and II. *J. appl. Phys.* **22** 57–64.

Brooker, M. P. & Morris, D. L. (1980) A survey of the macroinvertebrate riffle fauna of the rivers Ystwyth and Rheidol, Wales. *Freshwater Biology* **10** 459–474

Brown, A. W. A (1978) *Ecology of pesticides.* John Wiley & Sons, New York

Brown, B. E. (1976) Observations on the tolerance of the isopod *Asellus meridianus* Rac. to copper and lead. *Water Res.* **10** 555–559

Brown, B. E. (1977) Effects of mine drainage on the river Hayle, Cornwall. A. Factors affecting concentrations of copper, zinc and iron in water, sediments and dominant invertebrate fauna. *Hydrobiologia* **52** 221–232

Brown, V. M. (1969) The calculation of the acute toxicity of mixtures of poisons to rainbow trout. *Water Res.* **2** 723–733

Brown, V. M. (1973) Concepts and outlook in testing the toxicity of substances to fish. In: Glass, G. E. (ed) *Bioassay techniques and environmental chemistry.* Science Publishers, Ann Arbor

Brown, V. M. & Dalton, R. A. (1970) The acute lethal toxicity to rainbow trout of mixtures of copper, phenol, zinc and nickel. *J. Fish. Biol.* **2** 211–216

Brown, V. M., Jordan, D. H. M. & Tiller, B. A. (1967) The effect of temperature on the acute toxicity of phenol to rainbow trout in hard water. *Water Res.* **1** 587–594

Brown, V. M., Jordan, D. H. M. & Tiller, B. A. (1969) The acute toxicity to rainbow trout of fluctuating concentrations and mixtures of ammonia, phenol and zinc. *J. Fish. Biol.* **1** 1–9

Brown, V. M., Mitrovic, V. V. & Stark, G. T. C. (1968) Effects of chronic exposure to zinc on toxicity of a mixture of detergent and zinc. *Water Res.* **2** 255–63

Brown, V. M., Shaw, T. L. & Shurben, D. G. (1974) Aspects of water quality and the toxicity of copper to rainbow trout. *Water Res.* **8** 797–803

Brown, V. M., Shurben, D. G. & Fawell, S. K. (1967) The acute toxicity of phenol to rainbow trout in saline waters. *Water Res.* **1**, 683–685

Bryan, G. W. (1979) Bioaccumulation of marine pollutants. *Phil. Trans. R. Soc. Lond.* **B286** 483–505

Bryan, G. W. (1984) Pollution due to heavy metals and their compounds. In: Kinne, O. (ed) *Marine ecology* Vol V: *Ocean management Part 3: Pollution and Protection of the Seas.* John Wiley & Sons, Chichester

Buikema, A. L., McGinnis, M. J. & Cairns, J. (1979) Phenolics in aquatic ecosystems: a selected review of recent literature. *Mar. Environ. Res.* **2** 87–181

Burrows, W. D. (1977) Aquatic aluminium: chemistry, toxicology & environmental prevalence. *CRC Critical Reviews of Environmental Control* **7** 167–216

Burton, D. T., Morgan, E. L., & Cairns, J. (1972) Mortality curves of bluegills (*Lepomis macrochirus*) simultaneously exposed to temperature and zinc stress. *Trans. Amer. Fish. Soc.* **101** 435–441

Burton, J. D. (1979) Physico-chemical limitations in experimental investigations. *Phil. Trans. R. Soc. Lond.* **B286** 443–456

Cairns, J. (1974) Protozoans (Protozoa). *In*: Hart, C. W. & Fuller, S. H. (eds) *Pollution ecology of freshwater invertebrates.* pp. 1–25, Academic Press, New York

Cairns, J. (1979) A strategy for the use of protozoans in the evaluation of hazardous substances. In: James, A. & Evison, L. (eds) *Biological indicators of water quality.* John Wiley & Sons, Chichester

Cairns, J., Douglas, W. A., Busey, F. & Chaney, M. D. (1968) The sequential comparison index — a simplified method for non-biologists to estimate relative differences in biological diversity in stream pollution studies. *J. Wat. Poll. Contr. Fed.* **40** 1607–1613

Cairns, J., Heath, A. G. & Parker, B. C. (1975) The effects of temperature upon the toxicity of chemicals to aquatic organisms. *Hydrobiologia* **47** 135–171

Calamari, D. & Marchetti, R. (1973) The toxicity of mixtures of metals and surfactants to rainbow trout *Salmo gairdneri* (Rich.) *Water Res.* **7** 1453–1464

Calamari, D., Marchetti, R. & Vailati, G. (1980) Influence of water hardness on cadmium toxicity to *Salmo gairdneri*. *Water Res.* **14** 1421–1426

Carpenter, K. E. (1922) The fauna of the Clarach stream (Cardiganshire) and its tributaries. A preliminary study of the problem of lead pollution. *Aberystwyth Studies* **4** 251–258

Carpenter, K. E. (1924) A study of rivers polluted by lead mining in the Aberystwyth district of Cardiganshire. *Ann. appl. Biol.* **11** 1–23

Carpenter, K. E. (1925) On the biological factors involved in the destruction of river fisheries by pollution due to lead mining. *Ann. appl. Biol.* **12** 1–13

Carpenter, K. E. (1926) The lead mine as an active agent in river pollution. *Ann. appl. Biol.* **13** 395–401

Castenholz, R. W. & Wickstrom, C. E. (1975) Thermal streams. In: Whitton, B. A. (ed) *River ecology*. Blackwell Scientific Publications, Oxford, pp. 264–285

Carter, L. (1962) Bioassay of trade wastes. *Nature* **196** 2411

Chandler, J. R. (1970) A biological approach to water quality management. *Wat. Pollut. Control,* **69** 415–422

Chapman, G. A. (1978) Toxicities of cadmium, copper and zinc to four juvenile stages of chinook salmon and steelhead. *Trans. Am. Fish. Soc.* **107** 841–847

Chapman, P. M., Farrell, M. A. & Brinkhurst, R. D. (1982) Relative tolerances of selected aquatic oligochaetes to combinations of pollutants and environmental factors. *Aquatic Toxicology* **2** 69–78

Clark, R. B. (ed) (1982a) *The long-term effects of oil pollution on marine populations, communities and ecosystems*. The Royal Society, London. (Also published as Phil. Trans. Roy. Soc. Lond. *B297* 183–443, 1982.)

Clark, R. B (1982b) Environmental science and all that *Mar. Pollut. Bull.* **13** 335–336

Clark, R. B. (1986) *Marine pollution*. Clarendon Press, Oxford

Cole, H. A. (ed) (1979) *The assessment of sublethal effects of pollutants in the sea*. The Royal Society, London. (Also published as Phil. Trans. R. Soc. Lond. **B286** 399–633.)

Coleman, M. J. & Hynes, H. B. N. (1970) The vertical distribution of the invertebrates in the bed of a stream. *Limnol. Oceanogr.* **15** 31–40

Colwell, R. R. (ed) (1984) *Vibrios in the environment*. John Wiley & Sons, New York

Cook, S. E. (1976) Quest for an index of community structure sensitive to water pollution. *Environ. Pollut.* **11** 269–288

Curds, C. R. (1975) Protozoa. In: Curds, C. R. & Hawkes, H. A (eds) *Ecological aspects of used-water treatment, Vol. I. The organisms and their ecology*. Academic Press, London, 203–268

Curds, C. R. & Hawkes, H. A. (eds) (1975) *Ecological aspects of used-water treatment, Vol. 1. The Organisms and their Ecology*. Academic Press, London

Curds, C. R. & Hawkes, H. A. (eds) (1983a) *Ecological aspects of used-water treatment, Vol. II. Biological activities and treatment processes*. Academic Press, London

Curds, C. R. & Hawkes, H. A. (1983b) *Ecological Aspects of Used-water Treatment, Vol. III. The Processes and their Ecology*. Academic Press, London

Curtis, E. J. C. & Curds, C. R. (1971) Sewage fungus in rivers in the United

Kingdom: studies of *Sphaerotilus* slimes using laboratory recirculating channels. *Water Res.* **5** 267–279

Davenport, J. (1982) Oil and planktonic ecosystems. *Phil. Trans. R. Soc. Lond.* **B297** 369–384

Davies, R. P. & Dobbs, A. J. (1984) The prediction of bioconcentration in fish. *Water Res.* **18** 1253–1262

Davies, J. M. & Gamble, J. C. (1979) Experiments with large enclosed ecosystems. *Phil. Trans. R. Soc. Lond.* **B286** 523–544

Davies, P. H., Goettl, S. P. & Sinley, J. R. (1970) Toxicity of silver to rainbow trout (*Salmo gairdneri*) *Water Res.* **12** 113–117

Davis, J. C. (1973) Sublethal effects of bleached kraft pulp mill effluent on respiration and circulation in sockeye salmon (*Oncorhynchus nerka*). *J. Fish. Res. Bd. Can.* **30** 369–377

Department of the Environment (1986) *River quality in England and Wales 1985.* A report of the 1985 survey. Her Majesty's Stationery Office, London

Department of Scientific & Industrial Research (1935) *Survey of the River Tees: Part II. The Estuary, Chemical and Biological.* Water Pollution Research Board Technical Paper No. 5. Her Majesty's Stationery Office, London

Department of Scientific & Industrial Research (1964) *Effects of polluting discharges on the Thames estuary.* Water Pollution Research Technical Paper No. 11. Her Majesty's Stationery Office, London

Diamond, L. S. (1983) Lumen-dwelling protozoa. In: Jensen, J. B. (ed) *In vitro cultivation of protozoan parasites.* CRC Press, Boca Raton

Dobberkau, H.-J., Walter, R. & Logan, K. (1981) The recovery of viruses from water: methods and applications In: Goddard, M. & Butler, M. (eds) *Viruses and wastewater treatment.* Pergamon Press, Oxford

Doohan, M. (1975) Rotifera. In: Curds, C. R. & Hawkes, H. A. (eds) *Ecological aspects of used-water treatment, Vol. I. The Organisms and their ecology.* Academic Press, London, pp. 289–304

Duthie, J. R. (1977) The importance of sequential assessment in test programmes for estimating hazards to aquatic life. In: Mayer, F. L. & Hamelink, J. L. (eds) Aquatic toxicology and hazard evaluation, Special Technical Publication No 634 Washington, American Society for Testing and Materials

Dugan, P. R. (1972) *Biochemical ecology of water pollution.* Plenum Press, New York

Dybern, B. I., Ackefors, H. & Elmgren, R. (1976) Recommendations on methods for marine biological studies in the Baltic sea. *Balt. Mar. Biol. Publ.* **1** 1–98

Eden, G. E., Bailey, D. A. & Jones, K. (1977) Water re-use in the United Kingdom. In: Shuval, H. I. (ed) *Water renovation and re-use.* Academic Press, New York.

Edwards, R. W., Oborne, A. C., Brooker, M. P. & Sambrook, H. T. (1978) The behaviour and budgets of selected ions in the Wye catchment. *Verh. Int. Ver. Theor. Angew. Limn.* **20** 1418–1422

Eisler, R. (1965) Some effects of a synthetic detergent on estuarine fishes. *Trans. Am. Fish. Soc.* **94** 26–31

Eisler, R. (1970) *Acute toxicities of organochlorine and organophosphorus insecticides to estuarine fishes.* US Dept. of the Interior; Bureau of Sport, Fisheries and Wildlife Technical Report No. 46

Eisler, R. (1971) Cadmium poisoning in *Fundulus heteroclitus* (Pisces: Cyprinodon-tidae) and other marine organisms. *J. Fish. Res. Bd Can.* **28** 1225–1234

Elliot, J. M. (1977) *Some methods for the statistical analysis of samples of benthic invertebrates.* Freshwater Biological Association, Scientific Publication No. 25

Elliot, J. M. & Drake, C. M. (1981a) A comparative study of seven grabs used for sampling benthic macroinvertebrates. *Freshwater Biology* **11**, 99–120

Elliot, J. M. & Drake, C. M. (1981b) A comparative study of four dredges used for sampling benthic macroinvertebrates in rivers. *Freshwater Biology* **11** 245–262

Ernst, W. (1984) Pesticides and technical organic chemicals. In: Kinne, O. (ed) *Marine ecology Vol V: Ocean management, Part 4: pollution and protection of the seas.* John Wiley & Sons, Chichester

Finney, D. J. (1971) *Probit analysis.* 3rd Edition. Cambridge University Press

Fisher, R. A., Corbet, A. S. & Williams, C. B. (1943) The relation between the number of species and the number of individuals in a random sample of an animal population. *J. Anim. Ecol.* **12** 42–58

Flower, R. J. & Battarbee, R. N. (1983) Diatom evidence for recent acidification of two Scottish lochs. *Nature,* Lond. **305** 130–132

Fowler, S. W., LaRosa, Y., Unlu, B., Oregioni, J. P., Villeneuve, D. L., Fukai, R., Vallen, D. & Brisson, M. (1979) Heavy metals and chlorinated hydrocarbons in pelagic organisms from the open Mediterranean sea. *Journ. Etud. Pollut. C.I.E.S.M.* **4** 155–158

Fraser, P. (1984) Epidemiology and water quality. In: Lack, T. J. (ed) *Environmental protection: standards, compliance and costs.* Ellis Horwood, Chichester, in collaboration with the Water Research Centre

Furse, M. T., Wright, J. F. Armitage, P. D. & Moss, D. (1981) An appraisal of pond-net samples for biological monitoring of lotic macro-invertebrates *Water Res.* **15** 679–690

Gaddum, J. H. (1948) *Pharmacology,* 3rd edn. Oxford University Press

Gale, M. L., Wixon, B. G., Hardie, M. G. & Jennet, J. C. (1973) Aquatic organisms and heavy metals in Missouri's new lead belt. *Wat. Res. Bulletin* **9** 673–688

Garland, J. H. N. & Rolley, H. L. J. (1977) Studies of river water quality in the Lancashire Tame. *Wat. Pollut. Contr.* **76** 301–326

Gaufin, A. R., Jensen, L. D., Nebekev, A. V., Nelson, T. & Teel, R. W. (1965) The toxicity of ten inorganic insecticides to various aquatic invertebrates. *Wat. Sew. Wks. J.* July, pp. 276–279

Gerba, C. P. (1981) Virus survival in wastewater treatment. In: Goddard, M. & Butler, M. (eds) *Viruses and wastewater treatment.* Pergamon Press, Oxford

Gerba, C. P. (1983) Methods for recovering viruses from the water environment. In: Berg, G. (ed) *Viral pollution of the environment.* CRC Press, Boca Raton

Giddings, J. M. (1983) Microcosms for the assessment of chemical effects on the properties of aquatic ecosytems. In: *Hazard assessment of chemicals: current developments.* Vol. 2, Academic Press, New York pp. 46–89

Gilhooley, E. (1988) Studies into the effects of dissolved zinc on the rate of downstream drift of *Gammarus pulex* in laboratory channels. Unpublished M.Sc. thesis, Sunderland Polytechnic.

Gleason, H. A. (1922) On the relation between species and area. *Ecology* **3** 158

Goldberg, E. D., Bowen, V. T., Farnington, J. W., Harvey, G., Martin, J. H.,

Parker, P., Riseborough, R. W., Robertson, W., Schneider, E. & Gamble, E. (1978) Mussel watch. *Environ. Conserv.* **5** 101–125

Goodman, G. T. (1974) How do chemical substances affect the environment? *Proc. Roy. Soc. Lond.* **B185** 127–148

Grant, B. F. & Mehrle, P. M. (1983) Endrin toxicosis in rainbow trout. *J. Fish. Res. Bd Can.* **30**, 31–40

Gray, J. S. (1979) Pollution-induced changes in populations. *Phil. Trans. R. Soc. Lond.* **B286** 545–561

Gray, J. S. & Mirza, F. B. (1979) A possible method for the detection of pollution-induced disturbance on marine benthic communities. *Mar. Poll. Bull.* **10** 142–146

Gray, J. S. & Pearson, T. H. (1982) Objective selection of sensitive species indicative of pollution-induced change in benthic communities. 1. Comparative methodology. *Mar. Ecol. Prog. Ser.* **9** 111–119

Green, D. W. J. (1984) Ecological and toxicological studies on the invertebrate fauna of metalliferous streams. Unpublished Ph.D thesis, Sunderland Polytechnic.

Green, M. B. (1983) The Macrofauna of Sludge-drying Beds. In: Curds, C. R. & Hawkes, H. A. (eds) *Ecological aspects of used-water treatment, Vol. II. Biological activities and treatment processes.* Academic Press, London pp. 261–300.

Haines, T. A. (1981) Acidic precipitation and its consequences for aquatic ecosystems: a review. *Trans. Amer. Fish. Soc.* **110** 669–707

Hamelink, J. L. (1977) Current bioconcentration methods and theory. In: *Aquatic toxicology and hazard evaluation. Proceedings of 1st Annual Symposium on Aquatic Toxicology.* American Society for Testing and Materials, Philadelphia S.T.P. No. 634

Harding, J. P. C. & Whitton, B. A. (1981) Accumulation of zinc, cadmium and lead by field populations of *Lemanea*. *Water Res.* **15** 301–319

Hart, C. W. & Fuller, S. L. H. (eds) (1974) *Pollution ecology of freshwater invertebrates.* Academic Press, New York

Hartley, J. P. (1979) Biological monitoring of the seabed in the Forties oilfield. In: *Proceedings of a conference on ecological damage assessment, November, 1979. Arlington, Virginia.* Society of Petroleum Industry Biologists pp. 215–253

Hartley, J. P. (1982) Methods for monitoring offshore macrobenthos. *Mar. Pollut. Bull.* **13** 150–154

Haslam, S. M. (1982) A proposed method for monitoring river pollution using macrophytes. *Env. Technol. Letters* **3** 19–34

Hawkes, H. A. (1963) *The ecology of waste water treatment.* Pergamon Press, Oxford

Hawkes, H. A. (1975) River zonation and classification. In: Whitton, B. A. (ed) *River ecology.* Blackwell Scientific Publications, Oxford

Hawkes, H. A. (1983a) Activated sludge. In: Curds, C. R. & Hawkes, H. A. (eds) *Ecological aspects of used-water treatment, Vol II. Biological activities and treatment processes.* Academic Press, London pp. 77–162

Hawkes, H. A. (1983b) Stabilisation ponds. In: Curds, C. R. & Hawkes, H. A. (eds)

Ecology of used-water treatment, Vol. II. Biological activities and treatment processes. Academic Press, London pp. 163–218.

Hellawell, J. M. (1977) Biological surveillance and water quality monitoring. In: Alabaster, J. S. (ed) *Biological monitoring of inland fisheries.* Applied Science Publishers, London

Hellawell, J. M. (1978) *Biological surveillance of rivers — a biological monitoring handbook.* Water Research Centre

Hellawell, J. M. (1986) *Biological indicators of freshwater pollution and environmental management.* Elsevier Applied Science Publishers, London

Herbert, D. W. M., Alabaster, J. S., Dart, M. C., & Lloyd, R. (1961) The effect of China-clay wastes on trout streams. *Int. J. Air Wat. Pollut.* **5** 56–74

Herbert, D. W. M., Elkins, G. H. J., Mann, H. T. & Hemens, J. (1957) Toxicity of synthetic detergents to rainbow trout. *Wat. Waste Treat. J.* **6** 394–397

Herbert, D. W. M. & Merkens, J. C. (1961) The effect of suspended mineral solids on trout. *Int. J. Air Wat. Poll.* **5** 46–55

Herbert, D. W. M. & Shurben, D. G. (1964) The toxicity of fluoride to rainbow trout. *Water Waste Treatm. J.* **10** 141–142

Herbert, D. W. M. & Shurben, D. S. (1965) The susceptibility of salmonid fish to poisons under estuarine conditions. II—ammonium chloride. *Int. J. Air Wat. Poll.* **9** 89–91

Herbert, D. W. M. & Wakeford, A. C. (1964) The susceptibility of salmonid fish to poisons under estuarine conditions. I. Zinc sulphate. *Int. J. Air Wat. Poll.* **8** 251–256

Hermanutz, R. O., Mueller, L. H., & Kempfert, K. D. (1973) Captan toxicity to fathead minnows (*Pimephales promelas*) bluegills (*Lepomis macrochirus*) and brook trout (*Salvelinus fontinalis*). *J. Fish. Res. Bd Can.* **30** 1811–1817

Hewlett, P. S. & Plackett, R. L. (1979) *The interpretation of quantal responses in biology.* Edward Arnold, London

Higgins, I. J. & Burns, R. G. (1975) *The chemistry and microbiology of pollution.* Academic Press, London

Hirsch, E. (1963) Strukturelemente von Alkylbenzolsulfonaten und ihr Einfluß auf das Verhalten von Fischen. *Vom Wass.* **30** 249–259

Her Majesty's Stationery Office (HMSO) (1983a) *Acute toxicity testing with aquatic organisms 1981. Methods for examination of waters and associated materials.*

Her Majesty's Stationery Office (HMSO) (1983b) *Bacteriological examination of drinking water supplies 1982.*

Her Majesty's Stationery Office (HMSO) (1985) *Methods of biological sampling.* A colonisation sampler for collecting macro-invertebrate indicators of water quality in lowland rivers, 1983. Methods for the Examination of Waters and Associated Materials. 24pp.

Hokanson, K. E. F. & Smith, L. L. (1971) Some factors influencing the toxicity of linear alkylbenzene sulphonate (LAS) to the bluegill. *Trans. Amer. Fish. Soc.* **100** 1–12

Holme, N. A. & McIntyre, A. D. (eds) (1984) *Methods for the study of marine benthos,* 2nd ed. IBP Handbook No. 16: 350pp. Blackwell Scientific Publications, Oxford

Hornick, R. B. (1982) Typhoid fever. In: Evans, A. S. & Feldman, H. A. (eds) *Bacterial infections of humans.* Plenum Publishing Corporation, New York.

Howells, G. D. (1983) The effects of power station cooling water discharges on aquatic ecology. *Wat. Poll. Control* **82** 10–17

Howarth, R. S. & Sprague, J. B. (1978) Copper lethality to rainbow trout in waters of various hardness and pH. *Water Res.* **12** 455–462

Hunter, J. B., Ross, S. L. & Tannahill, J. (1980) Aluminium pollution and fish toxicity. *Wat. Poll. Control* **79** 413–420

Hurlbert, S. H. (1971) The nonconcept of species diversity: a critique and alternative parameters. *Ecology* **52** 577–586

Hynes, H. B. N. (1960) *The ecology of polluted waters.* Liverpool University Press

Hynes, H. B. N. (1970) *The ecology of running waters.* Liverpool University Press

James, A. & Evison, L. (1979) *Biological indicators of water quality.* John Wiley, Chichester

James M. Montgomery Inc. (1985) *Water treatment principles and design.* John Wiley & Sons, New York

Johnston, R. (1984) Oil Pollution and its Management. In: Kinne, O. (ed) *Marine ecology Vol. V: Ocean management. Part 4: Pollution and protection of the seas.* John Wiley & Sons, Chichester

Joint IMCO/FAO/UNESCO/WHO Group of Experts on the Scientific Aspects of Marine Pollution (1969). Abstract of the report of the first session. *Water Res.* **3** 995–1005

Jones, J. R. E. (1949) An ecological study of the river Rheidol, North Cardiganshire, Wales. *J. Anim. Ecol.* **18** 67–88

Jones, J. R. E. (1958) A further study of the zinc-polluted river Ystwyth. *J. Anim. Ecol.* **27** 1–14

Jones, J. R. E. (1940a) A study of the zinc-polluted river Ystwyth in North Cardiganshire, Wales. *Ann. appl. Biol.* **27** 367–378

Jones, J. R. E. (1940b) The fauna of the river Melindwr, a lead-polluted tributary of the river Rheidol in north Cardiganshire, Wales. *J. Anim. Ecol.* **9** 188–200

Kaesler, R. L., Herricks, E. E. & Crossman, J. S. (1978) Use of indices of diversity and hierarchical diversity in stream surveys. In: Dickson, K. L. & Cairns, J. (eds) Biological data in water pollution assessment: quantitative and statistical analysis. A.S.T.M. Special Technical Publication No. 652. American Society for Testing and Materials, Philadelphia

Kagi, J. H. R. & Nordberg, M. (eds) (1979) *Metallothionein.* Birkhauser Verlag, Basel

Keefe, T. J. & Bergersen, E. R. (1976) A simple diversity index based on the theory of runs. *Water Res.* **11** 689–691

Kendall, M. G. (1962) *Rank correlation methods.* Griffin and Co, London

Keusch, G. T. (1982) Shigellosis. In: Evans, A. S. & Feldman, H. A. (eds) *Bacterial infections of humans.* Plenum Publishing Corporation, New York

Khan, M. A. Q. (ed) (1977) *Pesticides in aquatic environments.* Plenum Press, New York

Kimerle, R. A. & Swisher, R. D. (1977) Reduction of toxicity of linear alkylbenzene sulphonate (LAS) by biodegradation. *Water Res.* **11** 31–37

Kinne, O. (ed) (1984a) *Marine ecology Volume V: ocean management. Part 3. Pollution and protection of the seas: radioactive materials, heavy metals and oil.* John Wiley & Sons, Chichester

Kinne, O. (ed) (1984b) *Marine ecology. Volume V: ocean management: Part 4. Pollution and protection of the seas: pesticides, domestic wastes and thermal deformations.* John Wiley & Sons, Chichester

Klein, L. (1966) *River pollution. 3. Control.* Butterworths, London

Kolkwitz, R. & Marsson, M. (1909) Okologie der Tierischen Saprobien. *Int. Rev. Ges. Hydrobiol.* **2** 125–152

Koryak, M., Shapiro, M. A. & Sykora, J. L. (1972) Riffle zoobenthos in streams receiving acid mine drainage. *Water Res.* **6** 1239–47

Kothe, P. (1962) Der "Artenfehlbetrag", ein Einfaches Gütekriterium und Siene Anwerdung bei Biologischen Vorfluteruntersuchungen. *Deutsche Gewasserkundl. Mitt.* **6** 60–65

Kumaguru, A. K. & Beamish, F. W. H. (1981) Lethal toxicity of Permethrin to rainbow trout *S. gairdneri,* in relation to body weight and temperature. *Water Res.* **15** 503–505

Lack, T. J. (ed) (1984) *Environmental protection: standards, compliance and costs.* Ellis Horwood, Chichester, for Water Research Centre.

Laurie, R. D. & Jones, J. R. E. (1938) The faunistic recovery of a lead-polluted river in north Cardiganshire, Wales. *J. Anim. Ecol.* **1** 272–289

Leach, J. M. & Thakore, A. N. (1973) Identification of the constituents of kraft pulping effluent that are toxic to juvenile Coho salmon (*Oncorhynchus kisutch*). *J. Fish. Res. Bd Canada* **30** 479–484

Lee, G. F. (1973) Chemical aspects of bioassay techniques for establishing water quality criteria. *Water Res.* **7** 1525–1546

Lemlin, J. S. (1980) The value of ecological monitoring in the management of petroleum industry discharges. *Water Sci. Technol.* **13** 437–464

Lester, W. F. (1975) Polluted river: River Trent, England. In: Whitton, B. A. (ed) *River ecology.* Blackwell Scientific Publications, Oxford; pp. 489–513

Letterman, R. D. & Mitsch, W. J. (1978) Impact of mine drainage on a mountain stream in Pennsylvania. *Environ Pollut.* 53–73

Lewis, J. R. (1972) Problems and approaches to baseline studies in coastal communities. In: Ruivo, M. (ed) *Marine pollution and sea life.* Fishing News Books for FAO, pp. 401–404.

Lindahl, P. & Cabridenc, R. (1978) Molecular structure-biological properties relationships in anionic surface-active agents. *Water Res.* **12** 25–30

Litchfield, J. T. (1949) A method for rapid graphic solution of time-per cent effect curves. *J. Pharmac. exp. Ther.* **97** 399–408

Litchfield, J. T. & Wilcoxon, F. (1949) A simplified method of evaluating dose-effect experiments. *J. Pharmac, exp. Ther.* **96** 99–113

Lloyd, R. (1960) Toxicity of zinc sulphate to rainbow trout. *Ann. appl. Biol.* **48** 84–94

Lloyd, R. (1961) Toxicity of mixtures of zinc and copper sulphates to rainbow trout (*Salmo gairdneri* Richardson). *Ann. appl. Biol.* **49** 535–538

Lloyd, R. (1965) Factors that affect the tolerance of fish to heavy metal poisoning.

In: *Biological problems in water pollution.* US Public Health Service, Washington, 99-WP-25, pp. 181–187

Lloyd, R. (1972) Problems in determining water quality criteria for freshwater fisheries. *Proc. R. Soc. Lond.* **B180** 429–449

Lloyd, R. & Orr, L. D. (1969) The diuretic response by rainbow trout to sublethal concentrations of ammonia. *Water Res.* **3** 335–344

Lundgren, D. G., Vestal, J. R. & Tabita, F. R. (1972) The Microbiology of Mine Drainage Pollution. In: Mitchell, R. (ed) *Water pollution microbiology.* Wiley Interscience, New York

Macek, K. J. & Sleight, B. H. (1977) Utility of toxicity tests with embryos and fry of fish in evaluating hazards associated with the chronic toxicity of chemicals to fishes. In: *Aquatic toxicology and hazard evaluation. Proceedings of the First International Symposium on Aquatic Toxicology. Philadelphia.* American Society for Testing and Materials, S.T.P. No. 634.

Maitland, P. S. (1978) *Biology of fresh waters.* Blackie & Sons, Glasgow

Maki, A. W. & Bishop, W. E. (1979) Acute toxicity studies of surfactants to *Daphnia magna* and *Daphnia pulex. Environ. Contam. Toxicol.* **8** 599–612

Maki, A. W. & Duthie, J. R. (1978) Summary of proposed procedures for the evaluation of aquatic hazard. In: *Estimating the hazard of chemical substances to aquatic life.* ASTM Special Technical Publication No. 657. Philadelphia: American Society for Testing and Materials.

Mance, G. (1987) *Pollution threat of heavy metals in aquatic environments.* Elsevier Applied Science, London

Marchetti, R. (1965) The toxicity of nonylphenol ethoxylate to the developmental stages of the rainbow trout. *S. gairdneri* Richardson. *Ann. appl. Biol.* **55** 425–430

Margalef, R. (1958) Information theory in ecology. *Gen. Syst.* **3** 36–71.

Marking, L. L. (1977) Method for assessing additive toxicity of chemical mixtures. In: Mayer, F. L. & Hamelink, J. L. (eds) *Aquatic Toxicology and Hazard Evaluation. Proceedings of the First International Symposium on Aquatic Toxicology,* Philadelphia: American Society for Testing and Materials, S.T.P. 634, pp. 99–108.

Mason, C. F. (1981) *Biology of freshwater pollution.* Longman, London

Mata, L. J. (1978) *The children of Santa Maria Cauque: a prospective field study of health and growth.* Cambridge (Mass): MIT Press.

Maugh, T. H. (1978) How many chemicals are there? *Science* **199** 162

McIntosh, R. P. (1967) An index of diversity and the relation of certain concepts to diversity. *Ecology* **48** 392–404.

McKim, J. M. (1977) Evaluation of tests with early life stages of fish for predicting long-term toxicity. *J. Fish. Res. Bd Can.* **34** 1148–1154

McLeay, D. J. (1976) A rapid method for measuring the acute toxicity of pulp mill effluents and other toxicants to salmonid fish at ambient room temperature. *J. Fish. Res. Bd. Canada* **33** 1303–1311

McLeay, D. J. & Brown, D. A. (1974) Growth stimulation and biochemical changes in juvenile coho salmo (*Oncorhynchus kisutch*) exposed to bleached kraft mill effluent for 200 days. *J. Fish. Res. Bd Can.* **31** 1043–1049

Mehrle, P. M. & Mayer, F. L. (1980) Clinical tests in aquatic toxicology: state of the art. *Environ. Health Perspectives* **34** 139–143

Menhinick, E. P. (1964) A comparison of some species-individuals diversity indices applied to samples of field insects. *Ecology* **45** 859–861

Ministry of Housing and Local Government (1969) Fish Toxicity Test. Her Majesty's Stationery Office, London

Moore, R. J. (1978) Is *Acanthaster planci* an r-strategist? *Nature* **271** 56–57

Moriarty, F. (1984) Persistant contaminants, compartmental models and concentrations along food chains. In: Rasmussen, S. (ed) *Ecotoxicology. Proceedings of the Third Oikos Conference Ecol. Bull.* (Stockholm) **36** 35–45

Mosey, F. E. (1983) Anaerobic processes. In: Curds, C. R. & Hawkes, H. (eds) *Ecology of used-water treatment, Vol. II. Biological activities and treatment processes.* Academic Press, London, pp. 219–260

Moss, B. (1980) *Ecology of fresh waters.* Blackwell Scientific Publications, Oxford

Mount, D. I. & Brungs, W. A. (1967) A simplified dosing apparatus for fish toxicology studies. *Wat. Res.* **1** 21–29

Mount, D. I. & Stephan, C. E. (1967a) A method for establishing acceptable toxicant limits for fish — malathion and butoxyethanol ester of 2, 4-D. *Trans. Am. Fish. Soc.* **96** 185–193

Mount, D. I. & Stephan, C. E. (1967b) Chronic toxicity of copper to the fathead minnow (*Pimephales promelas*) in soft water. *J. Fish. Res. Bd Can.* **26** 2449–2457

Mudrack, K. & Kunst, S. (1986) *Biology of sewage treatment and water pollution control.* Ellis Horwood, Chichester

Murphy, P. M. (1978) The temporal variability in biotic indices. *Environ. Pollut.* **17** 227–236

National Water Council (1981) *River quality: the 1980 survey and future outlook.* National Water Council, London

Norris, R. H., Lake, P. S. & Swain, R. (1982) Ecological effects of mine effluents on the South Esk river, Tasmania. III — Benthic macroinvertebrates. *Austr. J. Mar. Freshwat. Res.* **33** 789–809

Noy, J. & Feinmesser, A. (1977) The use of wastewater for agricultural irrigation. In: Shuval, H. I. (ed) *Water renovation and reuse.* Academic Press, New York

Nriagu, J. O. & Simmons, M. S. (1983) *Toxic contaminants in the Great Lakes.* John Wiley & Sons, New York

Nuttall, P. M. & Purves, J. B. (1974) Numerical indices applied to the results of a survey of the macroinvertebrate fauna of the Tamar catchment (south-west England). *Freshwater Biol.* **4** 213–222

Odonell, A. R., Mance, G. & Norton, R. (1984) *A review of the toxicity of aluminium in fresh water.* Technical Report No 197, Water Research Centre

Odum, H. T., Cantlon, J. E. & Kornicker, L. S. (1960) An organisational hierarchy postulate for the interpretation of species-individuals distribution, species entropy and ecosystem evolution, and the meaning of a species-variety index. *Ecology* **41** 395–399

Pagenkopf, G. K., Russo, R. C. & Thurston, R. V. (1974) Effect of complexation on toxicity of copper to fishes. *J. Fish. Res. Bd Can.* **31** 462–465

Pantle, R. & Buck, H. (1955) Die biologische uberwachung der Gewasser und die Darstellung der Ergebrisse. *Besondere Mitteilugen zum Deutschen Gewasserkundlichen Jahrbuch* **12** 135–143

Patten, B. C. (1962) Species diversity in net plankton of Raritan Bay. *J. Mar. Res.* **20** 57–75

Pearson, R. G. & Jones, N. V. (1975) The colonisation of artificial substrate by stream macroinvertebrates. *Progr. Wat. Technol.* **7** 497–504

Pearson, T. H., Gray, J. S. & Johannessen, P. J. (1983) Objective selection of sensitive species indicative of pollution-induced change in benthic communities. 2. Data analyses. *Mar. Ecol. Prog. Ser.* **12** 237–255

Perkins, E. J. (1974) *The biology of estuaries and coastal waters.* Academic Press, London

Perkins, E. J. (1979) The effects of marine discharges on the ecology of coastal waters. In: James, A. & Evison, L (eds) *Biological indicators of water quality.* John Wiley & Sons, Chichester

Perring, F. H. & Mellanby, K. (eds) (1977) *Ecological effects of pesticides. Linnean Society Symposium No. 5.* Academic Press, London

Phillips, J. (1972) Chemical processes in estuaries. In: Barnes, R. S. K. & Green, J. (eds) *The Estuarine Environment.* Applied Science Publishers, London

Pielou, E. C. (1984) *The interpretation of ecological data.* John Wiley & Sons, New York

Pitcairn, C. E. R. & Hawkes, H. A. (1973) The role of phosphorus in the growth of *Cladophora. Wat. Res.* **7** 159–71

Porter, E. (1973) *Pollution in four industrialised estuaries.* Her Majesty's Stationery Office, London

Preston, F. W. (1948) The commonness and the rarity of species. *Ecology* **29** 254–283

Raabe, E. W. (1952) Uber den "Affinititatswert" in der Planzensoziologie. *Vegetatio, Haag* **4** 53–68

Radford, D. S. & Hartland-Rowe, R. (1971) Subsurface and surface sampling of benthic invertebrates in two streams. *Limnol. Oceanogr.* **16** 114–120

Rafaelli, D. G. & Mason, C. F. (1981) Pollution monitoring with meiofauna, using the ratio of nematodes to copepods. *Mar. Poll. Bull.* **12** 158–163

Rahel, F. J. (1982) Population differences in acid tolerance between yellow perch, *Perca flavescens,* from naturally acidic and alkaline lakes. *Can. J. Zool.* **61** 147–152

Rao, V. C. & Melnick, J. L. (1986) *Environmental virology.* Van Nostrand Reinhold UK, Wokingham

Reiff, B., Lloyd, R., How, M. J., Brown, D. & Alabaster, J. S. (1979) The acute toxicity of eleven detergents to fish: results of an interlaboratory exercise. *Wat. Res.* **13** 207–210

Reish, D. J. (1973) The use of benthic animals in monitoring the marine environment. *J. environ. Plann. Pollut. Control* **1** 32–38

Reish, D. J. (1984) Domestic wastes. In: Kinne, O. (ed) *Marine ecology Vol V: ocean management part 4: pollution and protection of the seas.* John Wiley & Sons, Chichester

Roch, M., McCarter, J. A., Matheson, A. T., Clark, M. J. R. & Olafson, R. W. (1982) Hepatic metallothionein in rainbow trout (*Salmo gairdneri*) as an indicator of metal pollution in the Campbell river system. *Can. J. Fish. Aquat. Sci.* **39** 1596–1601

Royal Commission on Environmental Pollution (1972) *Third report: Pollution in some British estuaries and coastal waters.* Her Majesty's Stationery Office, London

Rye, R. P. & King, E. L. (1976) Acute toxic effects of two lampricides to twenty-one freshwater invertebrates. *Trans. Am. Fish. Soc.* **105** 322–326

Sanders, H. O. (1970) Toxicities of some herbicides to six species of freshwater crustaceans. *J. Wat. Poll. Contr. Fed.* **42** 1544–1550

Sattar, S. A. (1981) Virus survival in receiving waters. In: Goddard, M. & Butler, M. (eds) *Viruses and wastewater treatment.* Pergamon Press, Oxford

Say, P. J., Harding, J. P. C. & Whitton, B. A. (1981) Aquatic mosses as monitors of heavy metal contamination in the River Etherow, Great Britain. *Environ. Pollut. Series B* **2** 295–307

Schooner, J. L. (1982) Field validation of water quality criteria for hydrophobic pollutants. In: *Aquatic toxicology and hazard assessment. Proc. 5th Ann. Symp. on Aquatic Toxicology. Philadelphia.* American Society for Testing and Materials, S.T.P. No. 737.

Scullion, J. & Edwards, R. W. (1980) The effect of coal industry pollutants on the macroinvertebrate fauna of a small river in the South Wales coalfield. *Freshwater Biol.* **10** 141–162

Shannon, C. E. & Weaver, E. (1949) *The mathematical theory of communication.* University of Illinois Press, Urbana, pp. 82–83 and 104–107

Shaw, T. L. & Brown, V. M. (1974) The toxicity of some forms of copper to the rainbow trout. *Water Res.* **8** 377–382

Shiells, G. M. & Anderson, K. J. (1985) Pollution monitoring using the nematode-copepod ratio — a practical application. *Mar. Poll. Bull.* **16** 62–68

Shubert, E. (1984) *Algae as ecological indicators.* Academic Press, London

Shuval, H. I. (Ed) (1977a) *Water renovation and reuse.* Academic Press, New York

Shuval, H. I. (1977b) Health considerations in water reuse. In: Shuval, H. I. (ed) Water Renovation and Reuse. Academic Press, New York

Simpson, E. H. (1949) Measurement of diversity. *Nature* **163** 688

Sittig, M. (1975) *Environmental sources and emissions handbook.* New Jersey: Noyes Data Corporation.

Skalski, J. R. (1981) Statistical inconsistencies in the use of no-observed effect levels in toxicity testing. In: Branson, D. R. and Dickson, K. L. (eds) *Aquatic toxicology and hazard assessment. Proceedings Fourth Annual Symposium on Aquatic Toxicology,* American Society for Testing and Materials STP No. 737. Philadelphia

Skidmore, J. T. (1965) Resistance to zinc sulphate of the zebrafish (*Brachydanio rerio* Hamilton-Buchanan) at different phases of its life history. *Ann. appl. Biol.* **56** 47–53

Skidmore, J. F. (1970) Respiration and osmoregulation in rainbow trout with gills damaged by zinc sulphate. *J. Exp. Biol.* **52** 481–494

Skidmore, J. T. (1964) Toxicity of zinc compounds to aquatic animals, with special reference to fish. *Quart. Rev. Biol.* **39** 227–247

Slade, J. S. & Ford, B. J. (1983) Discharge to the environment of viruses in wastewater, sludges and aerosols. In: Berg, G. (ed) *Viral pollution of the environment*. CRC Press, Boca Raton

Sladecek, V. (1979) Continental systems for the assessment of river water quality. In: James, A. & Evison, L. (eds) *Biological indicators of water quality*. John Wiley, Chichester

Smith, J. E. (Ed) (1968) *"Torrey Canyon" pollution and marine life*. Cambridge University Press.

Smith, L. L., Broderius, S. J., Oseid, D. M., Kimball, G. L. & Koenst, W. M. (1978) Acute toxicity of hydrogen cyanide to freshwater fishes. *Arch. Environ. Contam. Toxicol.* **7** 325–337

Smith, L. L. & Oseid, D. M. (1972) Effects of hydrogen sulphide on fish eggs and fry. *Water Res.* **6** 711–720

Sneath, P. H. A. & Sokal, R. R. (1973) *Numerical taxonomy*. W. H. Freeman, San Francisco, pp. 141–145

Solbé, J. F. de L. G. (1973) The relation between water quality and the status of fish populations in Willow Brook. *Wat. Treatm. Exam.* **22** 41–61

Sorber, C. A. (1983) Removal of viruses from wastewater and effluents. In: Berg, G. (ed) *Viral pollution of the environment*. CRC Press, Boca Raton

Sorensen, T. (1948) A method of establishing groups of equal amplitude in plant sociology based on similarity of species content and its application to analyses of the vegetation on Danish commons. *Biol. Skr.* (K. denske indensk. Selske. N.S.) **5** 1–34

Southwood, T. R. E. (1978) *Ecological methods*. 2nd Edition. Chapman & Hall, London

Sprague, J. B. (1969) The measurement of pollutant toxicity to fish. I — Bioassay methods for acute toxicity. *Wat. Res.* **3** 793–821

Sprague, J. B. (1970) The measurement of pollutant toxicity to fish. II — Utilising and applying bioassay results. *Wat. Res.* **4** 3–32

Sprague, J. B. (1971) The measurement of pollutant toxicity to fish. III — Sublethal effects and 'safe' concentrations. *Water Res.* **5** 245–266

Sprague, J. B. (1973) The ABC's of pollutant bioassay using fish. In: *Biological methods for the assessment of water quality*. ASTM Special Technical Publication No 528 pp. 6–30. Philadelphia. American Society for Testing and Materials.

Sprague, J. B., Elson, P. F. & Saunders, R. L. (1965) Sublethal copper-zinc pollution in a salmon river — a field and laboratory study. *Int. J. Air Water Poll.* **9** 531–543

Sprague, J. B. & Ramsay, B. A. (1965) Lethal effects of mixed copper and zinc solutions for juvenile salmon. *J. Fish. Res. Bd Can.* **22** 425–432

Stebbing, A. R. D. (1979) An experimental approach to the determinants of biological water quality. *Phil. Trans. R. Soc. Lond.* **B286** 465–481

Steele, J. H. (1979) The uses of experimental ecosystems. *Phil. Trans. R. Soc. Lond.* **B286** 583–595

Stendahl, D. H. & Sprague, J. B. (1972) Effects of water hardness and pH on vanadium lethality to rainbow trout. *Water Res.* **16** 1479–1488

Stephan, C. E. (1977) Methods for calculating an LC50. In: Mayer, F. L. & Hamelink, J. L. (eds) *Aquatic toxicology and hazard evaluation. Proceedings of the First International Symposium on Aquatic Toxicology, Philadelphia.* American Society for Testing and Materials, S.T.P. No 634, pp. 65–84

Stiff, M. J. (Ed) (1980) *River pollution control.* Ellis Horwood, Chichester, for Water Research Centre.

Stirn, J. (1981) *Manual of methods in marine environmental research. Part 8. Ecological assessment of pollutant effects.* FAO Fish. Tech. Paper 209: 70 pp. FAO, Rome

Sutcliffe, D. W. & Carrick, T. R. (1973) Studies on mountain streams in the English Lake District. I. pH, calcium and the distribution of invertebrates in the River Duddon. *Freshwat. Biol.* **3** 437–462

Sutcliffe, D. W. (1983) Acid precipitation and its effects on aquatic systems in the English Lake District (Cumbria). *Freshwater Biological Association Annual Report No. 50,* pp. 30–62. Freshwater Biological Association, Ambleside

Swarts, F. A., Dunson, W. A., Wright, J. E. (1978) Genetic and environmental factors involved in increased resistance of brook trout to sulphuric acid solutions and mine-acid polluted waters. *Trans. Am. Fish. Soc.* **107** 651–677

Swedmark, M., Braaten, B., Emanuelsson, E. & Granmo, A. (1971) Biological effects of surface active agents on marine animals. *Mar. Biol.* **9** 183–201

Swisher, R. D., O'Rourke, J. T. and Tomlinson, H. D. (1964) Fish bioassays of linear alkylate sulphonate (LAS) and intermediate biodegradation products. *J. Amer. Oil-chem. Soc.* **41** 746–752

Thatcher, T. O. (1966) The comparative lethal toxicity of a mixture of hard ABS detergent products to eleven species of fishes. *Int. J. Air Wat. Poll,* **10** 585–590

Thatcher, T. O. & Santner, J. F. (1967) Acute toxicity of LAS to various fish species. *Proc. 21st Industrial Waste Conference,* Purdue University Engng. Ext. Ser. **121** 996–1002

Thorp, V. J. & Lake, P. S. (1973) Pollution of a Tasmanian river by mine effluents. II — Distribution of macroinvertebrates. *Int. Rev. Ges. Hydrobiol.* **58** 885–892

Thurston, R. V., Phillips, G. R., Russo, R. C. & Hinkins, S. M. (1981) Increased toxicity of ammonia to rainbow trout (*Salmo gairdneri*) resulting from reduced concentrations of dissolved oxygen. *Can. J. Fish. Aquat. Sci.* **38** 983–988

Tooby, T. E. (1978) A scheme for the evaluation of hazards to non-target aquatic organisms from the use of chemicals. *Proceedings of the Fifth Symposium of the European Weed Research Society,* pp. 287–294

Tsubaki, T. & Irukayama, K. (1977) *Minamata disease.* Elsevier, Amsterdam

Tyler, A. V. (1965) Some lethal temperature relations of two minnows of the genus *Chrosomus. Can. J. Zool.* **44** 349–364

Tyler, P. A. & Buckney, R. T. (1973) Pollution of a Tasmanian river by mine effluents. I — Chemical evidence. *Int. Rev. Ges. Hydrobiol.* **58** 873–883

United Nations Environment Programme (UNEP) (1985) *Mediterranean Action Plan.* Co-ordinating Unit for Oceans and Coastal Areas of the United Nations Environment Programme, Athens, 28 pp.

United Nations (1978) *Mediterranean Action Plan and the Final Act of the Conference of Plenipotentiaries of the Coastal States of the Mediterranean Region for the Protection of the Mediterranean Sea.* United Nations, New York

United Nations (1980) *Conference of Plenipoteniaries of the Coastal States of the*

Mediterranean Region for the Protection of the Mediterranean Sea against Pollution from Land-based Sources. Final Act and Protocol. United Nations, New York

United Nations (1984) *Protocol concerning Mediterranean Specially Protected Areas.* United Nations, New York

Varley, M. (1967) *British freshwater fishes: factors affecting their distribution.* Fishing News (Books), London

Vaughn, J. M. & Landry, E. F. (1983) Viruses in soils and groundwaters. In: Berg, G. (ed) *Viral pollution of the environment.* CRC Press, Boca Raton

Vigers, G. A. & Maynard, A. W. (1977) The residual oxygen bioassay: a rapid procedure to predict effluent toxicity to rainbow trout. *Wat. Res.* **11** 343–346

Washington, H. G. (1984) Diversity, biotic and similarity indices. A review with special reference to aquatic ecosystems. *Water Res.* **18** 653–694

Watton, A. J. & Hawkes, H. A. (1984) Studies on the effects of sewage effluent on gastropod populations in experimental streams. *Water Res.* **18** 1235–1248

Weatherley, A. H., Beavers, J. R. & Lake, P. S. (1967) The ecology of a zinc-polluted river. In: Weatherly, A. H. (ed) *Australian inland waters and their fauna: eleven studies.* Australian National University Press, Canberra, pp. 252–278

Wegl, R. (1983) Index fur die Limnosaprobitat. (Index for limnosaprobity). *Wasser und Abwasser* **26** 1–175

White, J. B. (1978) *Wastewater engineering.* Edward Arnold, London

Whittaker, R. H. (1965) Dominance and diversity in land plant communities. *Science* **147** 250–260

Whitton, B. A. (1970) Biology of *Cladophora* in freshwaters. *Water Res.* **4** 457–476

Whitton, B. A. (ed) (1975) *River ecology.* Blackwell Scientific Publications, Oxford

Whitton, B. A. (1979) Algae and higher plants as indicators of river pollution. In: James, A. & Evison, L. (eds) *Biological indicators of water quality.* John Wiley, Chichester

Whitton, B. A. (1980) Zinc and plants in rivers and streams. In: Nriagu, J. D. (ed) *Zinc in the environment. Part II. Health effects.* John Wiley, New York

Whitton, B. A. & Say, P. J. (1975) Heavy Metals. In: Whitton, B. A. (ed) *River ecology.* Blackwell Scientific Publications, Oxford, pp. 286–311

Whitton, B. A., Say, P. J. & Jupp, B. P. (1982) Accumulation of zinc, cadmium and lead by the aquatic liverwort *Scapania. Environ. Pollut. B.* **3** 299–316

Whitton, B. A., Say, P. J. & Wehr, J. D. (1981) Use of plants to monitor heavy metals in rivers. In: Say, P. J. and Whitton, B. A., (eds) *Heavy metals in Northern England:* environmental and biological aspects. University of Durham, Department of Botany.

Wildish, D. J. (1972) Acute toxicity of polyoxyethylene esters and polyoxyethylene ethers to *Salmo salar* and *Gammarus oceanicus. Wat. Res.* **6** 759–762

Wilhm, J. L. & Dorris, T. C. (1968) Biological parameters for water quality criteria. *Bioscience* **18** 477–481

Williams, C. B. (1964) *Patterns in the balance of nature, and related problems in quantitative ecology.* Academic Press, New York, pp. 14–31 and 147–192.

Williams, E. T. (1971) Principles of clustering. *Ann. Rev. Ecol. Syst.* **2**, 202–326

Winner, R. W. & Farrell, M. P. (1976) Acute and chronic toxicity of copper to four species of *Daphnia. J. Fish. Res. Bd Can.* **33** 1685–1691

Woodiwiss, F. (1964) The biological system of stream classification used by the Trent River Board. *Chemistry & Industry,* March 1964, **11** 443–447

Young, D. D. (1980) River pollution control by quality objectives. In: Stiff, M. J. (ed) *River pollution control,* Ellis Horwood, Chichester, for Water Research Centre

Zaman, V. & Ah Keong, L. (1982) *Handbook of medical parasitology.* ADIS Health Science Press, New York

Zahn, R. K., Zahn-Daimler, G., Muller, W. E. C., Michaelis, M. L., Kurelec, B., Rijavec, M., Batel, R. & Bihari, N. (1983) DNA damage by PAH and repair in a marine sponge. *Sci. Total. environ.* **26** 137–156

Index